Agricultural Science: Principles and Practices

Jessie Benson

SYRAWOOD
PUBLISHING HOUSE
New York

Published by Syrawood Publishing House,
750 Third Avenue, 9th Floor,
New York, NY 10017, USA
www.syrawoodpublishinghouse.com

Agricultural Science: Principles and Practices
Jessie Benson

International Standard Book Number: 978-1-64740-004-0 (Hardback)

Cataloging-in-Publication Data

Agricultural science : principles and practices / Jessie Benson.
p. cm.
Includes bibliographical references and index.
ISBN 978-1-64740-004-0
1. Agriculture. 2. Agronomy. 3. Crop science. 4. Agricultural resources.
I. Benson, Jessie.
S493 .A37 2020
630--dc23

Printed in China.

TABLE OF CONTENTS

Preface VII

Chapter 1 **Understanding Agricultural Science** 1

Chapter 2 **Agricultural Production and Productivity** 8

- Agricultural Product 8
- Crop 8
- Fruit Crop 15
- Vegetable Crop 25
- Cereals 36
- Wheat 38
- Rice 40
- Maize 51
- Sorghum 72
- Poultry Processing 73

Chapter 3 **Agricultural Processes and Resources** 80

- Crop Choice 80
- Soil Preparation 87
- Seed Selection and Sowing of Seeds 89
- Irrigation and Drainage 91
- Harvesting 98
- Agricultural Land 99
- Fertilizer 100
- Agricultural Machinery 111

Chapter 4 **Agricultural Soil Science** 114

- Soil Nutrients 120
- Soil Fertility 122
- Soil Texture 125

Chapter 5 **Pests and their Management** 130

- Plant Pests and Diseases 130
- Livestock Pests 149
- Pest Management 161

Chapter 6 **Crop Protection** **165**
- Crop Protection Methods 182
- Insect Management 188
- Plant Disease Management 203

Permissions

Index

PREFACE

Agriculture science is a biological field that is concerned with the practices and understanding of agriculture. It is an interdisciplinary subject which involves parts of natural, economic and social sciences. Agricultural science focuses on the research and development in the areas of plant pathology, horticulture, soil science, entomology, plant breeding and genetics, and production techniques such as irrigation management. It also deals with improving the quantity as well as the quality of agricultural productivity. Agricultural science also seeks to minimize the effects of pests such as weeds, insects and pathogens on the crop production system. This book provides comprehensive insights into the field of agricultural science. It is compiled in such a manner, that it will provide in-depth knowledge about the principles and practices in this area. This textbook is appropriate for those seeking detailed information about agricultural science.

To facilitate a deeper understanding of the contents of this book a short introduction of every chapter is written below:

Chapter 1- Agricultural science is a branch of biology that comprises of parts of social, natural, economic and exact sciences which are used in the practice and understanding of agriculture. It is concerned with plant breeding and genetics, horticulture, soil science, entomology, minimizing the effects of pests, etc. This is an introductory chapter which will introduce briefly all the significant aspects of agricultural science.

Chapter 2- Agricultural production is the process of obtaining various products by cultivating plants. The ratio of agricultural outputs to agricultural inputs is known as agricultural productivity. The topics elaborated in this chapter will help in gaining a better perspective about different agricultural products and their productivity.

Chapter 3- The various processes which are a part of agricultural production are known as agricultural processes. Some of these processes are crop choice, soil preparation, irrigation and drainage, and harvesting. This chapter discusses in detail the theories and methodologies related to these agricultural processes and resources.

Chapter 4- The study of edaphic conditions in connection with the production of food and fiber is known as agricultural soil science. It is primarily concerned with the study of soil nutrients, soil texture and soil fertility. The chapter closely examines these key concepts of agricultural soil science to provide an extensive understanding of the subject.

Chapter 5- A pest is any living organism which is harmful to plants, animals, humans, crops, livestock and forestry. Potato bug, locust and corn rootworms are some of the pests which affect agriculture. The different types of pests and how to manage them has been thoroughly discussed in this chapter.

Chapter 6- The science and practice of managing plant diseases, weeds and pests that damage crops and forestry is termed as crop protection. It is concerned with weed management, plant disease management and insect management. This chapter has been carefully written to provide an easy understanding of the varied facets of crop protection.

Finally, I would like to thank the entire team involved in the inception of this book for their valuable time and contribution. This book would not have been possible without their efforts. I would also like to thank my friends and family for their constant support

Jessie Benson

Chapter 1

Understanding Agricultural Science

Agricultural science is a branch of biology that comprises of parts of social, natural, economic and exact sciences which are used in the practice and understanding of agriculture. It is concerned with plant breeding and genetics, horticulture, soil science, entomology, minimizing the effects of pests, etc. This is an introductory chapter which will introduce briefly all the significant aspects of agricultural science.

Agricultural Science deals with the, food and fibre production and processing. They include the technologies of soil cultivation, crop cultivation and harvesting, animal production, and the processing of plant and animal products for human consumption and use.

Food is the most basic human need. The domestication and cultivation of plants and animals beginning almost 10,000 years ago were aimed at ensuring that this need was met, and then as now these activities also fit with the relentless human drive to understand and control the Earth's biosphere. Over the last century and a half, many of the world's political leaders have recognized what India's Jawaharlal Nehru did, that "Most things except agriculture can wait." Scientific methods have been applied widely, and the results have revolutionized agricultural production. Under the conditions of prescientific agriculture, in a good harvest year, six people can produce barely enough food for themselves and four others. Advanced technologies have made it possible for one farmer in the United States, for example, to produce food for more than 100 people. The farmer has been enabled to increase yields per acre and per animal; reduce losses from diseases, pests, and spoilage; and augment net production by improved processing methods.

Major Divisions

The agricultural sciences can be divided into six groups. In all fields, the general pattern of progress toward the solution of specific problems or the realization of opportunities is: (1) research to more accurately define the functional requirements to be served; (2) design and development of products, processes, and other means of better serving these requirements; and (3) extension of this information to introduce improved technologies to the agricultural industries. This has proved to be a tremendously successful approach and is being used the world over.

Soil and Water Sciences

Soil and water sciences deal with the geological generation of soil, soil and water physics and chemistry, and all other factors relevant to soil fertility. Soil science began with the formulation of the theory of humus in 1809. A generation later, Liebig introduced experimental science, including a theory of the supply of soil with mineral nutrients. In the 20th century, a general theory of soil fertility has developed, embracing soil cultivation, the enrichment of soil with humus and nutrients, and the preparation of soil in accordance with crop demands. Water regulation, principally drainage and irrigation, is also included.

Soil and water research have made possible the use of all classes of land in more effective ways, while the control of soil erosion and deterioration has made other advances even more striking. Because the amount of water available for plant growth is one of the major limiting factors in crop production, improved tillage and terracing practices have been devised to conserve soil moisture, and soil-management and land-use practices have been developed to increase the infiltration of snow, rain, and irrigation water, thereby reducing losses caused by runoff.

Public and private research into chemical fertilizers and soil management have made it possible for farmers to aid nature in making specific soils more productive. Much has been learned about using crop rotation, legumes, and green manure for replenishing soil humus and nitrogen; determining and supplying the major and minor nutrient needs of crops; and managing soil under irrigation, including salt control. Techniques based on these findings have been put to use on farms to improve soil fertility and increase crop yields. Between 1940 and 1965, for example, farmers in the United States more than tripled their use of chemical fertilizers, resulting in increases of 50 to 150 percent in crop yields.

Scientists have used many sophisticated techniques to unlock a vast storehouse of knowledge about plants. In one case, chemicals tagged with radioactive isotopes were employed to follow the processes by which plants take up soil nutrients to synthesize their fruits, grains, vegetables, nuts, flowers, and fibres.

Plant Sciences

The plant sciences include applied plant physiology, nutrition, ecology, breeding and genetics, pathology, and weed science, as well as crop management. They deal primarily with two major types of crops: (1) those that represent direct human food, such as cereals, vegetables, fruits, and nuts; and (2) those that serve as feed and forage for food, companion, laboratory, and recreational animals. Special branches of these sciences have developed to deal with each of the numerous classes of plant crops—e.g., vegetables, small fruits, citrus fruits and other tree fruits, and flowers and other ornamental plants. Other specialties concern the production of raw materials for industry—cotton, hemp, sisal, and silk—although some of these are losing economic importance in the face of competition from synthetic fibres. Branches of the plant sciences that deal with such tropical crops as coffee, tea, cocoa, bananas, coconuts, sugarcane, and pineapples, to the contrary, promise to retain their importance.

Although scientifically based plant production came of age at the end of the 19th century, it started much earlier. Instructions on sowing dates are reported in Egypt by 2000 BC. Throughout the centuries, numerous treatises have included recommendations on how to achieve higher and more efficient yields.

The stimulus for the development of the plant sciences did not come from botany but from agricultural chemistry, the application of which led to the development of plant physiology. Field experiments were started in Rothamsted, England, in 1834, and elsewhere in Europe soon after. Improved methods of experiment design and statistical analysis made possible the comparative study of plants and their cultivation systems.

Cultivation of plants by varieties had already led in the late 18th century to the systematic selection of cereal varieties according to predicted yield. The rediscovery at the start of the 20th century of Gregor Mendel's laws of heredity and later of ways to cause mutations led to modern plant breeding, with momentous results that included the tailoring of crop varieties for regions of climatic extremes. Agronomist Norman E. Borlaug was awarded the Nobel Prize for Peace for 1970 for the development of short-stemmed wheat, a key element in the so-called Green Revolution in developing countries.

Major advances in the study of plant diseases were recorded in the 19th century, and the science of plant nutrition matured in the second quarter of the 1900s. Serious calamities resulting from the introduction of plant diseases into regions where the indigenous plants had no immunity against them, and the invasion of grapes by insects and of potatoes by fungi, stimulated research efforts. During the 20th century, all diseases have become objects of systematic plant pathology research. Plant pathologists search for chemicals effective against microbial diseases, weeds, and various pests and seek to adjust the biotic balance to reduce losses. That chemical residues have created some problems has led to further scientific activity. Biological control measures may ultimately be less harmful to the environment and more specific and effective in pest and weed management.

Other research has been undertaken because consumers want better fruits and vegetables. New varieties have been developed, methods found to ensure that fresh and processed foods arrive at retail stores in prime condition, and grocers taught to care for these foods so that consumers receive them in the most attractive and nutritious state.

Animal Sciences

In modern civilizations, people rely on meat, milk, and eggs as major sources of numerous nutrients. To satisfy this demand, sheep, goats, cattle, water buffalo, swine, chickens, ducks, geese, and turkeys are produced on farms all over the world. To understand how agricultural animals convert feedstuffs into the food and other commodities consumers demand, animal scientists have undertaken broad investigations using highly sophisticated techniques. The animal sciences comprise applied animal physiology, nutrition, breeding and genetics, ecology and ethology, and livestock and poultry management. In addition, diseases of food animals are the focus of many veterinary scientists.

Animal nutrition research was well-established in several centres around the world by the turn of the 20th century, and it began to flourish during the second quarter of the 1900s. Many discoveries have been made about animal metabolism and consequent nutrient requirements; the usefulness of hundreds of feedstuffs as sources of essential amino acids, vitamins, and minerals, as well as lipids and carbohydrates; the proper balance of available nutrients in the diet; nutrient supplements and feed-processing technologies; and metabolite-partitioning and growth-promoting compounds. These fundamental findings have been applied widely since 1950, bringing about improved animal feeding. Studies of life processes in farm animals have helped in developing the optimal nutriment for each animal, and human nutrition has benefitted enormously from the knowledge that has come from these investigations.

The notion that "like begets like" was already current in biblical times. Long before the science of animal genetics developed, all species of agricultural animals were subjected to selective breeding to some extent. Modifying livestock and poultry to meet consumer demands requires the application of scientific principles to the selection of superior breeding animals and planned matings. For example, consumers have come to prefer more lean tissue and less fat in meat, and so the meat-type hog was developed in two decades of intensive selection and crossbreeding starting in the 1950s. Swine now yield more lean pork, grow faster, and require less feed to reach market weight than before. By the 1980s, a laying hen of any popular genetic strain, if managed properly, could be expected to produce more than 250 eggs annually, while special meat-producing strains of chickens gain body weight at a rate of 1 : 2 in ratio with feed intake.

Some of the most significant research in animal breeding has been done with dairy cattle and has established the proved sire system, in which bulls are ranked according to the performance of their offspring. The use of sires proved in this way together with artificial insemination has enabled dairymen to improve their herds by greatly expanding the influence of genetically superior bulls. Along with increased emphasis on performance testing, efforts have been made to predict at a young age whether an individual animal will be an efficient meat, milk, or egg producer. Such success has made for earlier culling and for herds and flocks of higher genetic merit.

Animals represent renewable agricultural resources because they reproduce, and animal scientists have studied animal reproduction assiduously since the 1930s. These investigations began in the United Kingdom but were soon joined by scientists in the United States, where the work blossomed. Basic discoveries have been put to use quickly in the animal industries. Elucidation of reproductive structures and mechanisms made it possible to refine reproductive management in the 1940s, and artificial insemination made possible the widespread use of proved sires in the 1950s. Additional basic knowledge and later technological developments made practical the control of the estrous cycle and of parturition by exogenous hormones and the serial harvesting and transplantation of embryos from donor females of high merit. The result of these changes has been an increase in the reproductive rate and efficiency of all species of farm animals.

Animal ecology and ethology are young branches of the animal sciences. Around the middle of the 20th century, environmental physiologists in the United States and the United Kingdom began to study agricultural animals' relations with their environment, including temperature, air, light, and diet. Interactions among environmental temperature, diet, and the animals' genetic makeup have been characterized, and great strides have been made in improving thermal-environmental management on farms. Lighting management is now essential to profitable poultry production, and the light environment is being controlled in livestock houses as well. Since the 1970s emphasis has shifted to include the behavioral adaptability of animals to their surroundings and the effects of environmental stress on the immune status of livestock and poultry. Farmers have widely adopted intensive systems of animal production, and these systems continue to present opportunities and problems to animal scientists concerned with discovering and accommodating the environmental and ethological needs of food animals.

Animal health is essential to the efficient production of wholesome animal products. An example of the economic effect of animal-disease research conducted by veterinary scientists is the control of Marek's disease, a highly contagious disease affecting the nerves and visceral organs of chickens, which resulted in a loss of more than $200,000,000 annually to the U.S. poultry industry alone. The disease was studied for more than 30 years before it was learned that it is caused by a herpes virus. Within three years of this discovery, a vaccine was developed that reduced the frequency of

Marek's disease and the resultant meat condemnations in vaccinated chickens by 90 percent and increased egg production by 4 percent. Veterinary scientists also investigate the chronic infectious diseases associated with high morbidity rates and various metabolic disorders.

Food Sciences and Other Post-harvest Technologies

A group of sciences and technologies underlie the processing, storage, distribution, and marketing of agricultural commodities and by-products. Modern post-harvest technology helps provide inexpensive and various food supplies for consumers, meets the demands of a variety of industrial users, and even creates replacements for fossil fuels.

Research having particular significance to post harvest technology includes genetic engineering techniques that increase the efficiency of various chemical and biological processes and fermentations for converting biomass to feedstock and for use in producing chemicals (including alcohols) that can replace petroleum-based products. Among the expected outcomes are the manufacture of new products from reconstituted ones and the recovery of by-products that would otherwise be considered waste.

Agricultural Engineering

Agricultural engineering includes appropriate areas of mechanical, electrical, environmental, and civil engineering, construction technology, hydraulics, and soil mechanics.

The use of mechanized power and machinery on the farm has increased greatly throughout the world, fourfold in the United States since 1930. Research in energy use, fluid power, machinery development, laser and microprocessor control for maintaining grain quality, and farm structures is expected to result in further gains in the efficiency with which food and fibre are produced and processed.

Agricultural production presents many engineering problems and opportunities. Agricultural operations—soil conservation and preparation; crop cultivation and harvesting; animal production; and commodities transportation, processing, packaging, and storage—are precision operations involving large tonnages, heavy power, and critical factors of time and place. Facilities designed to aid farm operations help farm workers to minimize the time and energy requirements of routine jobs.

Four primary branches have developed within agricultural engineering, based on the problems encountered. Farm power and machinery engineering is concerned with advances in farm mechanization—tractors, field machinery, and other mechanical equipment. Farm structures engineering studies the problems of providing shelter for animals and human beings, crop storage, and other special-purpose facilities. Soil and water control engineering deals with soil drainage, irrigation, conservation, hydrology, and flood control. Electric power and processing engineering is concerned with the distribution of electric power on the farm and its application to a variety of uses, such as lighting to control plant growth and certain animal production operations.

Agricultural Economics

The field of agricultural economics includes agricultural finance, policy, marketing, farm and agribusiness management, rural sociology, and agricultural law. The idea that the individual farm enterprise forms a unit—affected by location, production techniques, and market factors—originated during the 19th century. It was later supplemented by the theory of optimum utilization of production factors by the selection of production lines. Further refinement came about through applications of modern accounting methods. Research into farm and agribusiness management led to mathematical planning systems and statistical computation of farm-enterprise data, and interest has been drawn to decision-making behaviour studies of farm managers.

Agricultural policy is concerned with the relations between agriculture, economics, and society. Land ownership and the structure of farm enterprises were traditionally regarded as primarily social problems. The growth of agricultural production in the 20th century, accompanied by a decline in size of the rural population, however, has given impetus to research in agricultural policy. In the capitalist countries, this policy has concentrated on the influence of prices and market mechanisms; in the centrally planned countries, emphasis has been placed on artificially created market structures.

Research in agricultural marketing was originally limited to the problem of supply and demand, but the crises of the Great Depression in the 1930s brought new analytical studies. In Europe, the growth of the cooperative movement—begun in Germany in the 19th century as a response to capital shortage and farm indebtedness—brought satisfactory solutions to problems of distribution of products from farmer to processor. Consequently, little interest in market research developed in Europe until the mid-20th century. Today, agricultural marketing studies focus on statistical computations of past market trends to supply data for forecasting.

Agricultural law concentrates on legal issues of both theoretical and practical significance to agriculture such as land tenure, land tenancy, farm labour, farm management, and taxation. From its beginnings at the University of Illinois in the 1940s, modern agricultural law has evolved to become a distinct field of law practice and scholarship.

Rural sociology, a young discipline, involves a variety of research methods, including behaviour study developed from studies in decision making in farm management.

Other Agricultural Sciences

Agricultural work science arose in response to the rural social problems experienced in Germany during the Great Depression. The improvement of work procedures, appropriate use of labour, analysis of human capacity for work, and adjustment of mechanized production methods and labour requirements represent the main objects of this branch of ergonomics research. Studies of the influence of mechanization on the worker and of worker training came later.

Agricultural meteorology deals with the effects of weather events, and especially the effects of their variations in time and space, on plant and animal agriculture. Atmospheric factors such as cloud type and solar radiation, temperature, vapour pressure, and precipitation are of vital interest to agriculturalists. Agricultural meteorologists use weather and climatic data in enterprise risk analysis as well as in short- and long-range forecasting of crop yields and animal performance.

Emerging Agricultural Sciences

The agricultural sciences are poised to enter a new era, armed with ever more sophisticated research technologies, such as monoclonal antibodies and gene splicing, in their continuing drive to better harness nature for the ultimate benefit of human beings everywhere. Although broad and deep scientific investigations have been made in the biological, physical, and social realms related to agriculture, the need persists for additional research to close remaining gaps in knowledge, especially in molecular biology and the environmental, social, and economic effects of its fruits.

From results of experiments already conducted, it is clear that molecular biology will influence plant genetics and crop production. Plant genetic engineers are working to improve specific economically important plant varieties by increasing their photosynthetic efficiency, improving their nutritional quality, and transferring to them such favourable properties as the ability to fix atmospheric nitrogen, as do legumes, and to better resist diseases and tolerate herbicides and natural environmental stress.

Animal scientists also are using new research methods in biotechnology, including the micromanipulation of embryos to produce multiple clones. Monoclonal antibodies are used in studies of specific factors in immune mechanisms, and recombinant DNA (deoxyribonucleic acid) technology is used in the genetic engineering of microbes so that they can synthesize specific antigenic proteins useful in vaccine production. The ultimate goal of this research is to improve dramatically the health and productivity of agricultural animals.

Chapter 2

Agricultural Production and Productivity

Agricultural production is the process of obtaining various products by cultivating plants. The ratio of agricultural outputs to agricultural inputs is known as agricultural productivity. The topics elaborated in this chapter will help in gaining a better perspective about different agricultural products and their productivity.

Agricultural Product

Agricultural products are derived from cultivated plants or animals to sustain or enhance human life. Food is the most widely produced agricultural product, and, the global per-person food supply as measured in calories per person has risen more than 20 percent in the past 50 years.

Categories of Agricultural Products

Agricultural products fall into one of four groups: foods, fuels, fibers, and raw materials:

1. Food: Grains and cereal crops are grown on more than half the world's farmed acreage. But food agriculture crops encompass more than just cereals like wheat and corn. Meats and dairy products like milk also are agricultural food products, as are honey and farmed fish.

2. Fuels: Ethanol, produced from corn, sugarcane, or sorghum, is the agricultural fuel product in the widest use. However, agricultural byproducts like straw sugarcane also are burned to produce power.

3. Fibers: Fiber crops include cotton (one of the top 10 crops produced in the U.S. every year), wool, and silk. Agricultural producers also use hemp to make rope and flax for linen. It's even possible to use bamboo fiber to make cloth.

4. Raw Materials: These are agricultural products used to make other agricultural products. For example, livestock feed, considered an agricultural product, is used to provide nourishment to the animals that produce dairy products.

Crop

A crop is a plant or plant product that can be grown and harvested for profit or subsistence. By use, crops fall into six categories: food crops, feed crops, fiber crops, oil crops, ornamental crops, and industrial crops.

Food crops, such as fruit and vegetables, are harvested for human consumption. Grains, such as corn, wheat, and rice, are the world's most popular food crops.

Food crops were the first crops to be harvested through agriculture. Agricultural development and the growth of civilizations led to the diversity of other types of crops.

Feed Crops

Feed crops, such as oats and alfalfa, are harvested for livestock consumption. These crops contain nutrients that animals need to develop. They are grown in agricultural fields but can also be found in natural meadows and pastures.

The most popular type of feed crop is a forage crop. Animals feed directly on forages, such as grasses. Forages that are cut and fed to livestock while they are still fresh are called green chop. Alfalfa is a popular crop fed to livestock as green chop.

Some forages are cut, allowed to dry in the field, and stored. These are called hay crops.

Another type of forage crop is silage. Silage crops are harvested, then stored under conditions that allow the forage to break down (ferment) into acids. The wet, acidic silage is fed to livestock such as cattle.

Principle feed crops include corn, barley, wheat, and oats. Each of these crops has different properties that are better suited for some animals' diets over others. Barley, which is harder to digest, is most often fed to beef and dairy cattle because they have a tough, four-chambered stomach. Hull-less barley, which is easier to digest, is fed to swine and poultry.

The production of feed crops has risen dramatically with increased demand for meat worldwide. Increased production of feed crops has changed the agricultural landscape.

The Food and Agriculture Organization (FAO) says 33 percent of arable land on Earth is used to produce food for livestock. This limits the production of crops for human consumption, especially for the world's poorest people. Feed crops and grazing pastures disturb natural water cycles, drawing away water from underground aquifers that provide the earth with a constant and balanced supply of water.

Forests have been cleared to create pastures where livestock can graze. Almost 70 percent of land cleared from the Amazon rain forest, for instance, has been turned over to grazing.

Fiber Crops

Fiber crops, such as cotton and hemp, are harvested for textile and paper products. Textiles, or cloth, are made from the dried and processed fibers of certain plants. Most fibers used to make textiles are taken from the stem or roots of plants such as flax. Flax is used to make linen.

Other parts of a plant can be harvested for fiber. Cotton, the most popular fiber crop in the world, is harvested from the light, fluffy "boll" of fiber that surrounds the plant's seeds. Textiles made from bamboo are manufactured from the pulp of bamboo plants.

Pulp from other fiber crops can be used in a variety of products. Fiber pulp may be used instead of wood pulp to manufacture paper products.

The hemp plant is an interesting and controversial example of a fiber crop. The fibers of the hemp plant are strong and durable, perfect for products such as paper, textiles, ropes, nets, and sailcloth for ships. Hemp advocates see the plant as a versatile and ecological source of fiber.

But some varieties of the hemp plant are used to make marijuana, a narcotic drug. Marijuana is illegal to grow and use in most parts of the United States. (The drug is legally grown and sold for medical use in some places.) Opponents of hemp argue that increased harvesting of hemp crops will lead to increased production and use of marijuana.

Oil Crops

Oil crops, such as canola and corn, are harvested for consumption or industrial uses. Technologies developed in the past century have enabled crops to be processed and broken down into their primary components, including oil. Soybeans, for example, represented 56 percent of world oilseed production and 79 percent of all edible oil consumed in the United States in 2000.

Oil crops are harvested for use in cooking, such as olive oil and corn oil. Oil crops are also harvested for industrial use, such as oil paints, soaps, and lubrication for machinery.

Fuel made from oil crops is called biofuel. The demand for biofuels has grown in recent years. Rising gas prices, concerns about global warming, and a desire for energy self-sufficiency have led governments and businesses to invest in biofuel research.

There are two main types of biofuel that use oil crops: bioethanol and biodiesel.

Bioethanol is an alcohol made from fermented materials that come from sugar and starch crops. These crops include sugar cane, corn, and wheat. Bioethanol can be used as a fuel for vehicles, but it is usually used as a gasoline additive to improve vehicle emissions. Bioethanol is used widely in the United States and Brazil, where an abundance of corn and sugar cane crops facilitate its production.

Biodiesel is made by combining vegetable oils with alcohol. Nuts, such as coconuts, macadamias, and pecans, are excellent sources of oil used to manufacture biodiesel. Biodiesel can be used in diesel engines, such as those used by buses. Brazil, the United States, and the European Union (particularly Germany) manufacture and use biodiesel on a large scale.

Biofuels provide almost 2 percent of the world's transport fuel. Many scientists and economists predict that number will rise as oil production decreases in the next century.

Ornamental Crops

Ornamental crops, such as dogwood and azalea, are harvested for landscape gardening. Ornamental crops are most often grown in nurseries, where they are purchased for residential or commercial settings.

Ornamental crop production has deep historical roots. The tulip crop of the Netherlands, for example, has become a symbol of that country.

Today, ornamental crop production is an important economic activity in many developing countries. Kenya, for example, is a major exporter of roses and carnations. Kenyan flower growers have situated their greenhouses near the shores of Lake Naivasha and Lake Victoria, where the soil is fertile and the water is abundant and fresh.

Kenya's huge flower operations, however, are having a negative impact on lake ecosystems. Growers irrigate their flowers with lake water, dramatically lowering supplies of freshwater available for consumption and hygiene. Growers also apply heavy amounts of fertilizers and pesticides so their flowers can maintain their beauty throughout the export process. These chemicals often runoff into the lakes, endangering aquatic animal and plant life.

Industrial Crops

Industrial crops, such as rubber and tobacco, are harvested for their products' use in factories or machines. Industrial crops include all crops used in the production of industrial goods, such as fiber and fuel products.

Rubber is produced naturally from a wide variety of plants, but predominantly from the Hevea tree indigenous to Brazil. Rubber is harvested for its latex. Latex is an extremely tough fluid found in the inner bark of the Hevea tree. Latex is obtained by tapping—cutting or shaving the bark with a sharp knife—and collecting the latex in cups. When mixed with chemicals, latex creates solid rubber blobs, called curds. Rubber curds are pressed between rollers to remove excess moisture and to form sheets. The sheets are packed and shipped for use in tires, machine belts, shoe soles, and other products.

Rubber has been used by civilizations for thousands of years. One of the earliest uses of rubber was to create balls for use in games in the Olmec Empire in what is today Mexico. Today, rubber is still used to manufacture durable toys, as well as boots, flooring, balloons, and medical supplies.

Hevea trees transplanted to southern Asia now produce most of the world's rubber. The countries with the largest rubber crops are Thailand, Indonesia, and Malaysia. Industrialization around the world has increased the global demand for rubber. High demand for natural rubber has increased the environmental degradation of forests in southern Asia.

Harvesting Crops

Methods for growing and harvesting crops have developed over thousands of years. The earliest crops were grown in Mesopotamia around 5500 BCE. These crops, indigenous to an agriculturally rich area called the Fertile Crescent, were grown near local sources of freshwater so they could be irrigated relatively easily. Wheat, barley, and figs were among the first crops.

The development of agriculture led to more sophisticated methods of harvesting crops. Crop rotation was the most significant innovation. In crop rotation, one crop is planted one year, then a different crop is planted the next year on the same land. This helps preserve the soil and reduce the chance for disease.

Crop rotation and fertilization, which makes soil more productive, allowed farmers to grow more crops on less land. These innovations also allowed crops to be grown in areas where they might not grow naturally. Improved engineering allowed rivers to be dammed and diverted to provide water for crops. All of these developments increased the abundance of crops, which could be used for trade and industrial use.

Today, agriculture is the largest industry in the world. Millions of people harvest crops for subsistence or business purposes. Some tools used to harvest crops have not changed in a thousand years—plows, rakes, sickles. Most of all, harvesting crops still relies on human labor.

The tools and machinery used to harvest crops have grown much more complex and expensive, however. Fertilizers, which many farmers need to be economically competitive, cost more than many farmers in the developing world can afford. Machinery, such as tractors and plows, can cost hundreds of thousands of dollars.

GMOs

Technological and biological advancements in agriculture—such as chemical pesticides and fertilizers—have increased crop growth and resilience while they have decreased crop losses and prices.

Genetically modified organisms (GMOs or GM foods) are common throughout the developed world. Biotechnology allows scientists to alter the DNA of microbes, plants, and animals. Businesses sell farmers genetically modified seeds. With these seeds, farmers can use toxic chemicals without harming the crop. Farmers who grow GM foods increase production with less labor and less land. Vegetables and fruits last longer and are less likely to bruise.

The heavy reliance on chemicals has disturbed the natural environment, however. Helpful species of animals may be killed along with harmful ones. Chemical use may also pose a health hazard to people, especially through runoff entering local aquifers and other water supplies. Critics argue that GM foods have less nutritional value and decrease biodiversity.

Organic and free-range food industries have grown in opposition to industrial farming. Agricultural scientists are looking for safer chemicals to use as fertilizers and pesticides. Some farmers use natural controls and rely less on chemicals.

Seed Banks

In order to preserve biodiversity, seed banks have been created around the world to store seed samples. Seed banks may specialize in a specific crop or in the crops of a region. The International Potato Center, based in Lima, Peru, houses 150 wild potato species and other tubers of Andean origin.

Native Seeds, founded in the southwestern United States, helps Native Americans locate seeds for growing traditional crops, such as orach, or "mountain spinach," and amaranth, once widely used for food and fiber in Mexico.

The Svalbard Global Seed Vault, the world's most diverse seed bank, was established in 2008. The Norwegian government built the Seed Vault into the side of a permafrost-covered mountain on the island of Spitsbergen, part of the Svalbard archipelago about 1,030 kilometers (620 miles) from the North Pole. The vault is designed to safely store the seeds of hundreds of thousands of plant varieties from crops grown throughout the globe. The Seed Vault offers "fail-safe" protection for the world's agricultural inheritance against any natural, social, or economic disaster.

Today, the Seed Vault stores about 420,000 seed samples. It has the capacity to hold 4.5 million samples.

Crops have a wide variety of uses and are an integral part of our existence and development. While advancements in crop science and technology have increased the production of some of our most basic foodstuffs, they also have had wide-ranging impacts on the environment.

The production of crops does not have to harm the environment. By protecting the land, water, and air, and by sharing knowledge and resources, people may find solutions for the problems of world hunger and global energy scarcities through the sustainable use of crops.

Cash Crop

A cash crop is one that is cultivated to be sold to gain profit from the sale. The term helps to distinguish crops grown for commercial gain from the subsistence crops which are mainly grown to feed the farmer and his family with little left to be sold in the market. In the past, most crops were grown as food for the producer and his family. However, most of the crops grown today, especially in the developed world, are cash crops which are sold in the national or international markets to earn profits. Most of the cash crops cultivated in the developing nations are primarily exported to the developed economies.

Cash Crops of World's Eco-climatic Zones

Cash Crops of The Arctic: The large-scale cultivation of crops for commercial purpose is usually not possible in the harsh Arctic conditions. However, a few crops like the Rhodiola rosea, a medicinal herb grown here, is sold as a cash crop with its demand exceeding supply under the present conditions.

Cash Crops of The Temperate Zone: A large number of cash crops are grown in the temperate ecoregions of the world. These include cereals like wheat, corn, oats, etc.; vegetables like potatoes, fruits like cherries, apples, strawberries, etc., and oil-yielding crops like mustard, rapeseed, etc.

Subtropical Area's Cash Crops: Examples of some notable cash crops in the subtropical areas of the world include soybeans, a few herbs, etc.

Cash Crops of The Tropical Zone: A variety of crops like bananas, cocoa, coffee, sugar cane, cotton, jute, oranges, oil palm, etc., are grown in the tropical areas of the world to give high yields of these tropical crops.

Global Cash Crops

Coconut palms are cultivated in more than 80 countries of the world, with a total production of 61 million tonnes per year. The oil and milk derived from it are commonly used in cooking and frying; coconut oil is also widely used in soaps and cosmetics.

Sustainability of Cash Crops

Approximately 70% of the world's food is produced by 500 million smallholder farmers. For their livelihood they depend on the production of cash crops, basic commodities that are hard to differentiate in the market. The great majority (80%) of the world's farms measure 2 hectares or less. These smallholder farmers are mainly found in developing countries and are often unorganized, illiterate or enjoyed only basic education. Smallholder farmers have little bargaining power and incomes are low, leading to a situation in which they cannot invest much in upscaling their businesses. In general, farmers lack access to agricultural inputs and finance, and do not have enough knowledge on good agricultural and business practices. These high level problems are in many cases threatening the future of agricultural sectors and theories start evolving on how to secure a sustainable future for agriculture. Sustainable market transformations are initiated in which industry leaders work together in a pre-competitive environment to change market conditions. Sustainable intensification focuses on facilitating entrepreneurial farmers. To stimulate farm investment, projects on access to finance for agriculture are also popping up. One example is the SCOPE methodology, an assessment tool that measures the management maturity and professionalism of producer organizations as to give financing organizations better insights in the risks involved in financing. Currently agricultural finance is always considered risky and avoided by financial institutions.

Cash Crop Farming

Cash crop farming refers to a type of farming where agricultural crops are grown for the purpose of sale or to make profits, instead of subsistence or barter. It is also called commercial farming or cash cropping.

In simpler words, cash crop farming is done by farmers to earn money in return for sustenance or to meet the family's requirements. Also, an additional profit would be required for the next crop-related activity. That is, farmers have to borrow money from banks to purchase seeds for planting. Then, depending on the rise in market prices, they sell the harvested crops. In this way, they are able to repay their loans only after the sale of their crops.

The farming techniques used vary with each farmer as well as from one country to another. That is, traditional farmers in developing countries employ farming techniques that they are accustomed to. In the more developed countries, intensive cultivation and mechanized farming techniques are commonly used.

It has been observed that in the earlier times, cash crops were a small but critical part of the total production of a farm. Now, mostly in developed countries, the purpose of growing cash crops is earning revenue.

Furthermore, the price of cash crops depends on the supply and demand in more developed countries, the least developed countries being the suppliers.

Cash crop costs are majorly determined by the commodity markets with a global scope. However, there would be some fluctuations depending on the freight prices and local demand and supply.

The climate is a factor that decides the growth of cash crops. That is, the growth of cereals, fruit trees, and potatoes is supported by temperate climatic conditions, whereas rice, soybean, etc., come from a subtropical climate and sugarcane, cocoa, etc., from a tropical climate.

Advantages

- Cash crop farming is considered as an accurate method that has proven to raise affordable food in high quantities.

- It is profitable to the farmers and serves as a source of their sustenance.
- It gives employment where cash crops can be processed and promotes economic diversification.
- It earns revenue for the government.

Disadvantages

- Monocropping or sole cropping is followed, wherein a single crop is grown each year on the same land. Due to this, there may be limited production of certain food crops.
- Moreover, the continuous use of monocropping has been linked to soil degradation or decline in the soil quality, which further leads to the growth of pests and disease-causing pathogens. The outcome of this could also be mass starvation caused due to the extensive destruction of a particular crop.
- Cash crop farming may prove beneficial only to those farmers who have food security and access to other necessary inputs and income, whereas small farmers may face constraints.

Examples of Cash Crops

- Wheat, rye, corn, oats, barley, rapeseed, mustard, potatoes, rice, millet, apples, oranges, cherries, coffee, cotton, strawberries, raspberries, soy beans, tea, etc., are some common examples of cash crops.
- A well-known global cash crop is coconut and it is grown in over 80 countries having a climate suitable for its growth. Coconut and its derivatives are widely used in cooking, and in making soaps and cosmetics.
- Jatropha curcas is an example of a cash crop. It is used for the production of biofuel.
- Black market cash crops like coca, cannabis, and opium poppies are also produced.

Fruit Crop

Fruit is the fleshy or dry ripened ovary of a flowering plant, enclosing the seed or seeds. Thus, apricots, bananas, and grapes, as well as bean pods, corn grains, tomatoes, cucumbers, and (in their shells) acorns and almonds, are all technically fruits. Popularly, however, the term is restricted to the ripened ovaries that are sweet and either succulent or pulpy.

Botanically, a fruit is a mature ovary and its associated parts. It usually contains seeds, which have developed from the enclosed ovule after fertilization, although development without fertilization, called parthenocarpy, is known, for example, in bananas. Fertilization induces various changes in a flower: the anthers and stigma wither, the petals drop off, and the sepals may be shed or undergo modifications; the ovary enlarges, and the ovules develop into seeds, each containing an embryo plant. The principal purpose of the fruit is the protection and dissemination of the seed.

Fruits are important sources of dietary fibre, vitamins (especially vitamin C), and antioxidants. Although fresh fruits are subject to spoilage, their shelf life can be extended by refrigeration or by the removal of oxygen from their storage or packaging containers. Fruits can be processed into juices, jams, and jellies and preserved by dehydration, canning, fermentation, and pickling. Waxes, such as those from bayberries (wax myrtles), and vegetable ivory from the hard fruits of a South American palm species (Phytelephas macrocarpa) are important fruit-derived products. Various drugs come from fruits, such as morphine from the fruit of the opium poppy.

Types of Fruits

The concept of "fruit" is based on such an odd mixture of practical and theoretical considerations that it accommodates cases in which one flower gives rise to several fruits (larkspur) as well as cases in which several flowers cooperate in producing one fruit (mulberry). Pea and bean plants, exemplifying the simplest situation, show in each flower a single pistil (female structure), traditionally thought of as a megasporophyll or carpel. The carpel is believed to be the evolutionary product of an originally leaflike organ bearing ovules along its margin. This organ was somehow folded along the median line, with a meeting and coalescing of the margins of each half, the result being a miniature closed but hollow pod with one row of ovules along the suture. In many members of the rose and buttercup families, each flower contains a number of similar single-carpelled pistils, separate and distinct, which together represent what is known as an apocarpous gynoecium. In other cases, two to several carpels (still thought of as megasporophylls, although perhaps not always justifiably) are assumed to have fused to produce a single compound gynoecium (pistil), whose basal part, or ovary, may be uniloculate (with one cavity) or pluriloculate (with several compartments), depending on the method of carpel fusion.

Most fruits develop from a single pistil. A fruit resulting from the apocarpous gynoecium (several pistils) of a single flower may be referred to as an aggregate fruit. A multiple fruit represents the gynoecia of several flowers. When additional flower parts, such as the stem axis or floral tube, are retained or participate in fruit formation, as in the apple or strawberry, an accessory fruit results.

Certain plants, mostly cultivated varieties, spontaneously produce fruits in the absence of pollination and fertilization; such natural parthenocarpy leads to seedless fruits such as bananas, oranges, grapes, and cucumbers. Since 1934, seedless fruits of tomato, cucumber, peppers, holly, and others have been obtained for commercial use by administering plant growth substances, such as indoleacetic acid, indolebutyric acid, naphthalene acetic acid, and β-naphthoxyacetic acid, to the ovaries in flowers (induced parthenocarpy).

Classification systems for mature fruits take into account the number of carpels constituting the original ovary, dehiscence (opening) versus indehiscence, and dryness versus fleshiness. The properties of the ripened ovary wall, or pericarp, which may develop entirely or in part into fleshy, fibrous, or stony tissue, are important. Often three distinct pericarp layers can be identified: the outer (exocarp), the middle (mesocarp), and the inner (endocarp). All purely morphological systems are artificial. They ignore the fact that fruits can be understood only functionally and dynamically.

Classification of Fruits

There are two broad categories of fruits: fleshy fruits, in which the pericarp and accessory parts develop into succulent tissues, as in eggplants, oranges, and strawberries; and dry fruits, in which the entire pericarp becomes dry at maturity. Fleshy fruits include (1) the berries, such as tomatoes, blueberries, and cherries, in which the entire pericarp and the accessory parts are succulent tissue, (2) aggregate fruits, such as blackberries and strawberries, which form from a single flower with many pistils, each of which develops into fruitlets, and (3) multiple fruits, such as pineapples and mulberries, which develop from the mature ovaries of an entire inflorescence. Dry fruits include the legumes, cereal grains, capsulate fruits, and nuts.

As strikingly exemplified by the word nut, popular terms often do not properly describe the botanical nature of certain fruits. A Brazil nut, for example, is a thick-walled seed enclosed in a likewise thick-walled capsule along with several sister seeds. A coconut is a drupe (a stony-seeded fruit) with a fibrous outer part. A walnut is a drupe in which the pericarp has differentiated into a fleshy outer husk and an inner hard "shell"; the "meat" represents the seed—two large convoluted cotyledons, a minute epicotyl and hypocotyl, and a thin papery seed coat. A peanut is an indehiscent legume fruit. An almond is a drupe "stone"; i.e., the hardened endocarp usually contains a single seed. Botanically speaking, blackberries and raspberries are not true berries but aggregates of tiny drupes. A juniper "berry" is not a fruit at all but the cone of a gymnosperm. A mulberry is a multiple fruit made up of small nutlets surrounded by fleshy sepals. And strawberry represents a much-swollen receptacle (the tip of the flower stalk bearing the flower parts) bearing on its convex surface an aggregation of tiny brown achenes (small single-seeded fruits).

Fruit Farming

Fruit farming is the growing of fruit crops, including nuts, primarily for use as human food.

The subject of fruit and nut production deals with intensive culture of perennial plants, the fruits of which have economic significance (a nut is a fruit, botanically). It is one part of the broad subject of horticulture, which also encompasses vegetable growing and production of ornamentals and flowers. This topic places further arbitrary limitations in that it does not encompass a number of very important perennial fruit crops covered elsewhere, including vanilla, coffee, and the oil-producing tung tree and oil palm.

Botanists define a fruit in broad terms as the fleshy or dry ripened ovary surrounding the seed of a plant. A pomologist, or specialist in the science and practice of fruit growing, defines it somewhat more narrowly as the fleshy edible part of a perennial plant associated with development of the flower. A nut is any seed or fruit consisting of a kernel, usually oily, surrounded by a hard or brittle shell. Most edible nuts—e.g., almond, walnut, cashew, pecan, pistachio, etc.—are well known as dessert nuts. Not all nuts are edible. Some, used as sources of oil or fat, may be regarded as oil seeds; others are used for ornament. The botanical definition of a nut, based on features of form and structure (morphology), is more restrictive: a hard, dry, one-celled, one-seeded fruit that does not split open at maturity. Among the nuts that fit both the botanical and popular conception are the acorn, chestnut, and filbert; other so-called nuts may be botanically a seed (Brazil nut), a legume (peanut [groundnut]), or a drupe (almond and coconut).

Improvements in technology and consolidation of the fruit and nut industries in the most favoured climates of the world have been responsible for a steady increase in yield. Thus, the total acreage or number of plants devoted to various fruit and nut crops has dropped, remained about the same, or not risen in proportion to the increase in the respective crop production.

Although fruit- and nut-growing enterprises cover great ranges of climates and plant materials, their technologies have many common problems and practices. The most significant of these are discussed below.

The Variety: Propagation and Improvement

The first step in establishing a fruit- or nut-growing industry is the selection of individual plants with high productivity and a superior product. Such an individual is a horticultural variety. If it is multiplied vegetatively from rooted cuttings, from root pieces that throw shoots, or by graftage, each plant in the group (called a clone) that results is identical with the others. Nearly all commercially important perennial fruit and nut crops are clonally propagated; i.e., their varieties are multiplied vegetatively by one means or another. Some nut crops, such as the wild pecan, cashew, black walnut, hickory, and chestnut still come from trees that grow at random from seed; hence, character and quality tend to vary.

Many important varieties of fruit plants were selected generations ago. The Sultanina (Thompson Seedless) grape, the Lob Injir (Calimyrna) fig, and the Gros Michel banana have obscure origins; planted by the millions since selection, each specimen is actually a vegetative continuation of the selected individual growing on an independent root system. But regardless of the age of a fruit-growing industry, or the perfection of some of the selected varieties, a continuing search for new varieties is essential. There is always room for improvement in climatic adaptability, in insect and disease resistance, and in the solution of special horticultural or marketing problems. In fact, government experiment stations over the world now stress scientific breeding for improvement of market quality and yield of key fruit and nut crops.

Gros Michel banana trees.

Not only are varietal selection and improvement a continuing need but so also is the maintenance of existing varieties. Although an improved vegetative mutation of a variety is exceptional, the opportunities for accidental multiplication of degenerate (low-quality) mutants increase in proportion to the number of specimens of the variety. As a result, care is taken to propagate a clone only from superior individuals, and in the case of citrus, where mutation is especially common, further precautions are necessary. There are, of course, occasional mutations that may greatly improve a variety and these are sought, selected, and propagated.

Vegetative propagation technique varies with the individual fruit plant. Date, banana, and pineapple are multiplied by use of offshoots or suckers. Grape, fig, olive, currant, and blueberry are usually propagated from cuttings. Strawberry and black raspberry reproduce vegetatively by special organs—the former by stolons or runners, the latter by cane tip rooting or layering. Many kinds of fruit trees must be grafted or budded on especially grown rootstocks because the species to be multiplied does not root itself easily; apple, pear, peach, mango, and citrus are examples of this group. Many nut trees have a single taproot with but few branching roots, necessitating a deep hole and special care in transplanting.

Today's trend is toward a smaller tree in most fruit crops, particularly the apple and pear, and toward closer planting in hedgerow style, with carefully regulated fertilization and irrigation. This increases production per acre, lowers labour cost, increases early yields, and facilitates access in maintenance and harvesting. This approach, in fact, has been used for decades in Europe. Labour is the largest element of cost in fruit and nut production. Every means is exploited to reduce, facilitate, or eliminate hand labour.

With most fruit species a period of one to two years intervenes between the time a cutting is rooted and the time the plant is ready for setting in the field, or between graftage or budding and field planting. During this interval the plants remain in a nursery where they can be given intensive culture in rows. Pineapple and banana planting materials, however, do not require nursery care before field planting.

In choosing fruit varieties, the grower must (1) recognize the relative adaptabilities of available varieties to the climatic and soil conditions of his farm and (2) select a group that satisfies both his management needs and the market demands from those best adapted to his conditions. For instance, an apple producer in the northeastern U.S. may raise four varieties: Milton, McIntosh Red, Red Delicious, and Rome Beauty. The main harvest seasons for these succeed each other at two-week intervals; this helps him extend the harvest period and make efficient use of his labour. The first two varieties cross-fertilize satisfactorily, as do the last two. The first of these varieties is usually marketed without storage, while the storage seasons of the others are of increasing length. This helps the grower to extend his marketing period.

Cultivation

Site Selection

The site of a fruit-growing enterprise is as significant in determining its success as the varieties grown. In fact, variety and site together set a ceiling on the productivity and profit that can be realized under the best management. In most developed fruit regions microclimatic conditions (climate at plant height, as influenced by slight differences in soil, soil covering, and elevation) and soil conditions are the two components of a site that determine its desirability for a fruit-growing enterprise. Sometimes (particularly with highly perishable fruits) transportation to market must also be considered.

Local conditions at a site that expose it to unusual frost hazard are as detrimental to citrus in Florida as they are to peach trees in New Zealand and apple trees in the south of England. In regions and sites where temperatures during the season may drop no more than a few degrees below freezing, artificial frost protection is sometimes used. This is accomplished by open-flame burning (petroleum bricks, logs, etc.) or heating of metal objects with oil, gas, propane, electricity, etc. (stones or stacks that radiate heat). Another technique is the spraying of water on plants (e.g., strawberries) as long as the temperature is below freezing.

For highest productivity, most fruit trees must root extensively to a depth of three feet (one metre) or more. Heavy subsoil or other conditions causing imperfect internal drainage may result in shallow, weak root systems that do not take water and nutrients efficiently from the soil. In semi-arid and arid regions, accumulation of saline soils in a subsurface layer sometimes limits rooting of fruit trees, causes abnormal foliar symptoms, and reduces yields. Tiling and surface ditching help decrease water accumulation in poorly drained subsoils and reduce wet spots in otherwise satisfactory sites. Special control of irrigation procedures and periodic leaching may alleviate the worst salt effects in saline soils. Choice of tolerant species, varieties, and rootstocks may make fruit growing economical on imperfectly drained or mildly saline sites, though plants rarely perform as well as they do on sites free from these difficulties. Coconuts, however, tolerate saline soil conditions near tropical saltwater coasts.

Once selected, a site is cleared, levelled (if needed), and cultivated. Then drainage, irrigation, and road systems are installed as required. In rolling or sloping terrain, where contour planting is needed to control erosion and conserve moisture, the locations of the plant or row positions are determined by the contour terraces and waterways established. In old lands, nematode or other pest populations make fumigation necessary before planting. In some problem California soils, giant plows and treaded tractors turn the soil to depths of three to six feet (one to two metres). In very infertile sites, or sites where the physical condition of the surface soil is poor, it may be helpful to grow a succession of leguminous cover crops for a year or more before planting and/or apply a fertilizer containing major fertilizer elements (nitrogen, potassium, phosphorus, calcium, sulfur) and all or certain trace elements (iron, manganese, boron, zinc, copper, molybdenum) and lime, based on a soil test.

Planting and Spacing Systems

Growth, flowering habits, and light requirements on the one hand, and management problems on the other, determine the most satisfactory planting plan for a fruit- and nut-growing enterprise. There is a trend toward use of dwarfing stocks, growth control chemicals, or closer planting and training, or all of them to get the highest yields and best operation efficiency possible on a unit of ground.

Low-growing crops such as strawberry and pineapple are usually managed in beds containing several rows, or in less formal matted rows. In an acre of strawberries, 200,000 or more plants may occupy the matted rows. A pineapple plantation with two-row beds, having plants one foot (0.3 metre) apart in rows two feet (0.6 metre) apart totals 15,000 to 18,000 plants per acre (37,000 to 44,000 per hectare). With such dense populations, intense competition for light, water, and nutrients causes smaller average fruit size. Nevertheless, the total yield per unit of land is usually greater than it would be with lower plant numbers.

The spacing of grapevines along a trellis row and of trees planted in hedgerows involves the same group of problems. Maximum vineyard production frequently results with vine distances of eight to nine feet (2.4 to 2.7 metres; 600 ± per acre [1,500 per hectare]). The trend for peach trees and spur-type apple strains is hedgerows 14 feet (4.2 metres) apart or closer, in rows 18 to 20 feet (5.4 to 6 metres) apart.

With those species and varieties that require cross-pollination by insects, the planting plan must take those special needs into account. This is a problem with apple, pear, plum, and sweet cherry orchards. At least two varieties that cross-fertilize successfully must be planted in association with each other.

Training and Pruning

Pruning is the removal of parts of a plant to influence growth and fruitfulness. It is an important fruit-growing practice. Primary attention is given to form in the first few years after fruit trees or vines are planted. Form influences strength and longevity of the mature plant as well as efficiency of other fruit-growing practices; pruning for form is called training. As the plant approaches maximum fruitfulness and fills its allotted space, maintenance pruning for various purposes becomes increasingly important.

The grape may be trained following one of two systems: (1) spur system, cutting growth of the previous season (canes) to short spurs, (2) long-cane system, permitting canes to remain relatively long. Whether a spur or long-cane system is followed depends on the flowering habit of the variety. Relatively small trees that respond favourably to severe annual pruning (e.g., the peach and Kadota fig) are usually trained to create an open-centred tree with a scaffold of four or five main branches that originate on a short trunk and branch a number of times to provide fruiting wood. Annual renewal pruning can be reasonably efficient under these circumstances. Larger trees that do not respond favourably to heavy annual pruning are trained best to a system that encourages the main leader branch to grow erect to a height of eight to 10 feet (2.4–3.0 metres), with four or five main lateral branches at intervals on its sides forming the scaffold that carries fruiting wood up and out; this is called a modified leader system. The central leader type of tree, with one main leader up through the centre and many side branches, is common for pear and apple planted in hedgerows, and possibly for other fruits and nuts as the close-planted hedgerow system is more widely adopted.

The principal reasons for maintenance pruning are: (1) to permit efficient spraying and harvesting operations, (2) to maintain satisfactory light exposure for most of the leaves, and (3) to create a satisfactory balance between flowering and leaf surface.

To reduce hand labour costs, larger commercial fruit growers use machine pruning on many types of fruits. Peach, apple, pear, and other fruits usually planted in hedgerows are mowed across the top and sides by machine, then thinned out as needed by a follow-up crew using pneumatic clippers and hand-powered saws, operating from hydraulically manipulated scaffolds or lifts of various types.

Crop Enhancement

Pollination

The stimulus of pollination, fertilization, and seed formation is needed to get good size, shape, and flavour of most of the fruits. (Banana, pineapple, and some citrus and fig varieties are exceptions.) Transfer of pollen from the anthers (male) to the stigmas (female) is accomplished in nature either by insects or by movement in air. It is common practice to bring beehives into the orchard during bloom. Rainy cold weather during bloom with little or no sunshine can deter activity of the honey bee (the key insect pollinator) and reduce fruit set appreciably. This is one of the main problems not fully solved by fruit researchers. Hand-pollination by daubing collected and preserved pollen onto the stigma (as is done with date palms) sometimes is practiced for other fruits, but this approach is not widespread.

Thinning

Removal of flowers or young fruit (thinning) is done to permit the remaining fruits to grow more rapidly and to prevent development of such a large crop that the plant is unable to flower and set a commercial crop the following year. Thinning is done by hand, mechanically, or chemically. With the date, the pistillate flower cluster is reduced in size at the time of hand-pollination. In the case of certain table grape varieties, some clusters are cut off. With the Thompson seedless grape, a combination of girdling the trunk bark and judicious application of gibberellin (growth regulating) sprays at blossoming gives excellent full bunches.

Young peach fruits are thinned by striking the branches with a padded pole or by shaking the entire tree for a few seconds with a well-padded motor-driven shaker arm grasping the trunk. Hand thinning of young apple and peach fruits once was also a common practice, but because of the expense and difficulty, there has been increasing use of chemical sprays as a substitute. Two kinds of sprays are used: (1) mildly caustic sprays applied during bloom, such as Elgetol in arid regions, or (2) sprays of growth-regulating substances such as 3-CPA (2,3-chlorophenoxy propionamide) applied within a few weeks after bloom in areas with late frosts.

Pest Control and Preservation

In many fruit enterprises, pest control is the most expensive and time-consuming growing practice. Where the concentration of fruit farms in an area warrants it, individual efforts are complemented by legislative measures including quarantine regulations to force removal of pest-laden, unattended orchards. Sometimes the most economical control procedure is biological in nature. There is increased research today to find and multiply parasites that kill fruit crop pests. Such biological methods are necessary as political pressures increase for banning DDT and other chemicals. Selection of varieties that are immune, resistant to attack, or tolerant to specific pests, is a biological control procedure also widely used. Chemical control procedures, however, are relied on most heavily. Air-blast spray or mist-application machinery covering 70 acres (28 hectares) of trees or more in a day is now in common use.

Harvesting and Packing

The proper time to remove a fruit from the tree or plant varies with each fruit and is governed by whether the product will be sold and consumed within hours, or stored for weeks, months, or even a year. Most fruits are harvested as close as possible to the time they are eaten. A few, of which banana and pear are outstanding examples, may be harvested while immature and still ripen satisfactorily. Orange, grapefruit, and some varieties of avocado may be "stored" on the tree for several months after they have attained good quality; this method cuts costs in handling and marketing.

Many fruits, including apple, pear, orange, lemon, and grapefruit, may drop from the tree during the last part of the maturation period. Preharvest drop of these fruits can be delayed by application of dilute sprays of growth-regulating substances like naphthaleneacetic acid (NAA). The chemical spray Alar [N-(dimethylamino) succinamic acid] applied four to six weeks after bloom on apple not only reduces fruit drop at harvest but increases red colour, firmness, and return bloom the next year, in addition to other advantages.

For the fresh market, most tree and bush fruits are still harvested by hand. For processing, drying, and occasionally for fresh market, mechanical motor-driven tree and bush shakers with appropriate catching belts, bins, pallets, and electric lifts reduce harvesting and handling labour. In years to come, machinery may make it possible to machine-harvest most fruits, with no more, and possibly less, damage than with hand picking.

The public has become increasingly particular about the appearance and quality of the product it buys. Hence, store managers and suppliers seek the best grades of fruits and nuts available, and growers make every effort to produce crops with attractive colour and smooth finish. Fruits are packed by government-controlled grades such as Fancy or Extra Fancy within given size limits and are so labelled on the carton or box, together with the source. Most fruits and nuts not meeting this standard of quality are processed or sent through channels using the lower grades and off sizes.

Small packages of plastic foam or wood pulp base holding four to six fruits covered and heat sealed with polyethylene plastic film are popular. These are delivered to stores in corrugated cartons holding a few dozen packages. Citrus, apples, and whole nuts or kernels also are packaged in polyethylene bags and delivered in cartons. Loose fruit may be sold in cell cartons and tray packs consisting of stacked form-fitting pulp trays in a “bushel size” box. Every effort is made to eliminate bruising.

Large truck-pulled containers with individually motor-driven refrigeration units, with or without controlled atmosphere (CO_2-O_2, to retard ripening), are loaded at the fruit source and trucked to their destination or are loaded on ships by derrick for overseas shipment. These sealed containers are also being used increasingly for bananas to reduce labour and handling and to deliver the product in better condition.

Air shipment of “vine- and tree-ripe” fruit (strawberries, figs, sweet cherries, pineapples, avocados) to distances as far as from California to Europe in a day or less is becoming increasingly common with the much larger and faster cargo planes and reduced air-freight prices.

Postharvest Physiology of Fruits

Fruit ripening is a form of senescence and signifies the final stage in fruit development. A fleshy fruit is the enlarged ovary of a flower (avocado) or additional floral parts such as in apple, pear, and pineapple. Usually fertilization, and sometimes pollination alone, stimulate the floral parts causing a rapid cell division that leads to differentiation and the formation of the fruit structure. During this stage fruits consist of small, young cells filled with protoplasm. When the young fruit has been stimulated, presumably by plant hormones that originate from the embryonic seeds, rapid cell expansion takes place. During this stage fruits gain rapidly in size and weight. The cells develop small cavities or spaces in their tissue (become vacuolated) and begin the process of foodstuff accumulation, which lends fruits their compositional diversity. Banana, apple, and date, for example, accumulate mainly carbohydrates. Avocado and olive store fatty materials. Important constituents of most fruits are organic acids such as malic acid, found in apple and pear; citric acid, found in citrus fruits and pineapple; and tartaric acid, found in grapes. Fruits are usually low in protein.

After cell expansion has slowed and become nominal, fruits enter the stage of maturity and undergo preparation for ripening. Some crops, such as pear and avocado, are harvested at the so-called mature-green state and allowed to ripen afterward. Most fruits are at a stage of incipient ripening before they are picked. Ripening is marked by rapid and dramatic changes that give fruits their attractive and edible character. Some of the familiar changes are softening, which results from degradation of cell wall substances; disappearance of a green background, because of chlorophyll degradation (as in pear, apple, and banana); appearance of coloured pigments such as the carotenoids—orange-yellow—and anthocyanins—red (as in orange, mango, and strawberry); a decrease in acidity and increase in the sugar content (orange, apple); and emission of the volatile substances that give many fruits their distinct aroma (as in banana, pear, and apple). In climacteric fruits (e.g., banana, pear, apple), ripening is accompanied by increased respiration. In nonclimacteric fruit (e.g., strawberry, cherry) this phenomenon does not occur.

It is thought that the transition from the mature to the ripe stage is brought about by certain "ripening" enzymes. Protein molecules act as catalysts. The activity of these enzymes leads first to various ripening reactions, and then to gradual deterioration of the fruit tissue.

Because ripening leads to tissue breakdown, fruits are considered a highly perishable commodity. Different fruits have varying degrees of postharvest longevity. While strawberries last only a week to 10 days, for example, apples or lemons can be stored successfully for as long as several months.

Postharvest life of fruits can be extended by refrigeration with or without a modified oxygen–carbon dioxide atmosphere. Most temperate-zone fruits can be held safely at 32 to 41 °F (0 to 5 °C), but many subtropical and tropical fruits, including lemon, avocado, banana, and mango, show signs of injury from being chilled in prolonged cold storage and consequently fail to ripen properly. Bananas do not tolerate temperatures below 53 °F (12 °C), while several avocado varieties can be stored at temperatures as low as 46 °F (8 °C).

Fruit life can be extended further by both refrigeration and controlled atmosphere (CA) storage in which oxygen is kept at about 5 percent and carbon dioxide at 1 to 3 percent, while temperature is held at a level best suited to the particular fruit. So-called CA storage is common today for apples and pears and is being adapted to other fruits. Controlled atmosphere and refrigeration in conjunction with the removal of ethylene gas (which emanates from fruits and speeds ripening) helps slow the ripening process considerably. Golden Delicious apples and some pears are shipped in polyethylene containers in which a desirable, modified atmosphere is created by the respiring fruit.

Drying is a standard practice for stabilizing the market movement of dates, figs, raisin grapes, prunes, and apricots. Canning is of paramount importance to the pineapple, peach, and pear industries (these fruits can be dried as well), and freezing is a means of stabilizing some of the most perishable fruits, including strawberry, raspberry, and blueberry.

Nuts are susceptible to mold, souring, staleness, discoloration, and rancidity. Cured and dried nuts are kept in prolonged cold storage under controlled temperature and humidity levels. Nuts also are stored and sold in vacuum packs of carbon dioxide-enriched atmosphere.

Vegetable Crop

Vegetables, are mainly annual plants cultivated as field and garden crops in the open and under glass, and used almost exclusively for food. Vegetables grown principally for animal feed or seed should be excluded. Certain plants, normally classified as cereals and pulses, belong to this group when harvested green, such as green maize, green peas, etc. This grouping differs from international trade classifications for vegetables in that it includes melons and watermelons, which are normally considered to be fruit crops. But, whereas fruit crops are virtually all permanent crops, melons and watermelons are similar to vegetables in that they are temporary crops. Chillies and green peppers are included in this grouping when they are harvested for consumption as vegetables and not processed into spices.

Based on their edible botanical parts, plants are grouped into the following types of vegetables:

- Root Vegetables: Plants which are sources of edible roots, mainly modified underground roots, e.g., beets and sweet potato. These modified roots are either tuberous roots or fleshy roots.
- Stem Vegetables: Plants which are sources of edible botanical stems which are usually immature and succulent. They can be divided further into those plants in which the edible plant structure consists mainly of aerial stems, e.g., bamboo shoot and those with underground modified stems such as corms, tubers, bulbs, rhizomes, and stolons.
- Leafy Vegetables: Plants which are sources of edible leaves in various stages of development. The leaves may be separate and fully expanded or, as in cabbage, form a head. It includes, in some species, succulent stems, e.g., sweet potato (camote tops).
- Flower Vegetables: Plants which are sources of edible flowers or inflorescences including, in some species, accessory parts like the stalk (pedicel or peduncle). In cauliflower, it consists of the stalk and an immature inflorescence. But in squash, it is the mature flower.
- Fruit Vegetables: Plants which are sources of edible botanical fruits, either simple or compound, e.g., tomato and corn (maize).
- Seed Vegetables: Plants which are sources of edible botanical seeds, mostly under the family Leguminosae or Fabaceae (legumes), e.g., mung bean and pigeon pea.

Vegetable Farming

Vegetable farming is the process of growing of vegetable crops, primarily for use as human food.

The term vegetable in its broadest sense refers to any kind of plant life or plant product; in the narrower sense, however, it refers to the fresh, edible portion of a herbaceous plant consumed in either raw or cooked form. The edible portion may be a root, such as rutabaga, beet, carrot, and sweet potato; a tuber or storage stem, such as potato and taro; the stem, as in asparagus and kohlrabi; a bud, such as brussels sprouts; a bulb, such as onion and garlic; a petiole or leafstalk, such as celery and rhubarb; a leaf, such as cabbage, lettuce, parsley, spinach, and chive; an immature flower, such as cauliflower, broccoli, and artichoke; a seed, such as pea and lima bean; the immature fruit, such as eggplant, cucumber, and sweet corn (maize); or the mature fruit, such as tomato and pepper.

The popular distinction between vegetable and fruit is difficult to uphold. In general, those plants or plant parts that are usually consumed with the main course of a meal are popularly regarded as vegetables, while those mainly used as desserts are considered fruits. Thus, cucumber and tomato, botanically fruits, since they are the portion of the plant containing seeds, are commonly regarded as vegetables.

Types of Production

Vegetable production operations range from small patches of crops, producing a few vegetables for family use or marketing, to the great, highly organized and mechanized farms common in the most technologically advanced countries.

In technologically developed countries the three main types of vegetable farming are based on production of vegetables for the fresh market, for canning, freezing, dehydration, and pickling, and to obtain seeds for planting.

Production for the Fresh Market

This type of vegetable farming is normally divided into home gardening, market gardening, truck farming, and vegetable forcing.

Home gardening provides vegetables exclusively for family use. About one-fourth of an acre (one-tenth of a hectare) of land is required to supply a family of six. The most suitable vegetables are those producing a large yield per unit of area. Bean, cabbage, carrot, leek, lettuce, onion, parsley, pea, pepper, radish, spinach, and tomato are desirable home garden crops.

Market gardening produces assorted vegetables for a local market. The development of good roads and of motor trucks has rapidly extended available markets; the market gardener, no longer forced to confine his operations to his local market, often is able to specialize in the production of a few, rather than an assortment, of vegetables; a transformation that provides the basis for a distinction between market and truck gardening in the mid-20th century. Truck gardens produce specific vegetables in relatively large quantities for distant markets.

In the method known as forcing, vegetables are produced out of their normal season of outdoor production under forcing structures that admit light and induce favourable environmental conditions for plant growth. Greenhouses, cold frames, and hotbeds are common structures used. Hydroponics, sometimes called soilless culture, allows the grower to practice automatic watering and fertilizing, thus reducing the cost of labour. To successfully compete with other fresh market producers, greenhouse vegetable growers must either produce crops when the outdoor supply is limited or produce quality products commanding premium prices.

Production for Processing

Processed vegetables include canned, frozen, dehydrated, and pickled products. The cost of production per unit area of land and per ton is usually less for processing crops than for the same crops grown for market because raw material appearance is not a major quality factor in processing. This difference allows lower land value, less hand labour, and lower handling cost. Although many kinds of vegetables can be processed, there are marked varietal differences within each species in adaptability to a given method.

Specifications for vegetables for canning and freezing usually include small size, high quality, and uniformity. For many kinds of vegetables, a series of varieties having different dates of maturity is required to ensure a constant supply of raw material, thus enabling the factory to operate with an even flow of input over a long period. Acceptable processed vegetables should have a taste, odour, and appearance comparable with the fresh product, retain nutritive values, and have good storage stability.

Vegetables Raised for Seed Production

This type of vegetable farming requires special skills and techniques. The crop is not ready for harvest when the edible portion of the plant reaches the stage of maturity; it must be carried through further stages of growth. Production under isolated conditions ensures the purity of seed yield. Special techniques are applied during the stage of flowering and seed development and also in harvesting and threshing the seeds.

Production Factors and Techniques

Profitable vegetable farming requires attention to all production operations, including insect, disease, and weed control and efficient marketing. The kind of vegetable grown is mainly determined by consumer demands, which can be defined in terms of variety, size, tenderness, flavour, freshness, and type of pack. Effective management involves the adoption of techniques resulting in a steady flow of the desired amount of produce over the whole of the natural growing season of the crop. Many vegetables can be grown throughout the year in some climates, although yield per acre for a given kind of vegetable varies according to the growing season and region where the crop is produced.

Climate

Climate involves the temperature, moisture, daylight, and wind conditions of a specific region. Climatic factors strongly affect all stages and processes of plant growth.

Temperature

Temperature requirements are based on the minimum, optimum, and maximum temperatures during both day and night throughout the period of plant growth. Requirements vary according to the type and variety of the specific crop. Based on their optimum temperature ranges, vegetables may be classed as cool-season or warm-season types. Cool-season vegetables thrive in areas where the mean daily temperature does not rise above 70 °F (21 °C). This group includes the artichoke, beet, broccoli, brussels sprouts, cabbage, carrot, cauliflower, celery, garlic, leek, lettuce, onion, parsley, pea, potato, radish, spinach, and turnip. Warm-season vegetables, requiring mean daily temperature of 70 °F or above, are intolerant of frost. These include the bean, cucumber, eggplant, lima bean, okra, muskmelon, pepper, squash, sweet corn (maize), sweet potato, tomato, and watermelon.

Premature seeding, or bolting, is an undesirable condition that is sometimes seen in fields of cabbage, celery, lettuce, onion, and spinach. The condition occurs when the plant goes into the seeding stage before the edible portion reaches a marketable size. Bolting is attributed to either extremely low or high temperature conditions in combination with inherited traits. Specific vegetable strains or varieties may exhibit significant differences in their tendency to bolt.

Young cabbage or onion plants of relatively large size may bolt upon exposure to low temperatures near 50° to 55 °F (10° to 13 °C). At high temperatures of 70° to 80 °F (21° to 27 °C) lettuce plants do not form heads and will show premature seeding. The fruit sets of tomatoes are adversely affected by relatively low and relatively high temperatures. Tomato breeders, however, have developed several new varieties, some setting fruits at a temperature as low as 40 °F (4 °C) and others at a temperature as high as 90 °F (32 °C).

Moisture

The amount and annual distribution of rainfall in a region, especially during certain periods of development, affects local crops. Irrigation may be required to compensate for insufficient rainfall. For optimum growth and development, plants require soil that supplies water as well as nutrients dissolved in water. Root growth determines the extent of a plant's ability to absorb water and nutrients, and in dry soil root growth is greatly retarded. Extremely wet soil also retards root growth by restricting aeration. Atmospheric humidity, the moisture content of the air, also contributes moisture. Certain seacoast areas characterized by high humidity are considered especially adapted to the production of such crops as the artichoke and lima bean. High humidity, however, also creates conditions favourable for the development of certain plant diseases.

Daylight

Light is the source of energy for plants. The response of plants to light is dependent upon light intensity, quality, and daily duration, or photoperiod. The seasonal variation in day length affects the growth and flowering of certain vegetable crops. Continuation of vegetative growth, rather than early flower formation, is desirable in such crops as spinach and lettuce. When planted very late in the spring, these crops tend to produce flowers and seeds during the long days of summer before they attain sufficient vegetative growth to produce maximum yields. The minimum photoperiod required for formation of bulbs in garlic and onion plants differs among varieties, and local day length is a determining factor in the selection of varieties.

Each of the climatic factors affects plant growth, and can be a limiting factor in plant development. Unless each factor is of optimum quantity or quality, plants do not achieve maximum growth. In addition to the importance of individual climatic factors, the interrelationship of all environmental factors affects growth.

Certain combinations may exert specific effects. Lettuce usually forms a seedstalk during the long days of summer, but the appearance of flowers may be delayed, or even prevented, by relatively low temperature. An unfavourable temperature combined with unfavourable moisture conditions may cause the dropping of the buds, flowers, and small fruits of the pepper, reducing the crop yield. Desirable areas for muskmelon production are characterized by low humidity combined with high temperature. In the production of seeds of many kinds of vegetables, absence of rain, or relatively light rainfall, and low humidity during ripening, harvesting, and curing of the seeds are very important.

Site

The choice of a site involves such factors as soil and climatic region. In addition, with the continued trend toward specialization and mechanization, relatively large areas are required for commercial production, and adequate water supply and transportation facilities are essential. Topography—that is, the surface of the soil and its relation to other areas—influences efficiency of operation. In modern mechanized farming, large, relatively level fields allow for lower operating costs. Power equipment may be used to modify topography, but the cost of such land renovation may be prohibitive. The amount of slope influences the type of culture possible. Fields with a moderate slope should be contoured, a process that may involve added expense for the building of terraces and diversion ditches. The direction of a slope may influence the maturation time of a crop or may result in drought, winter injury, or wind damage. A level site is generally most desirable, although a slight slope may assist drainage. Exposed sites are not suitable for vegetable farming because of the risk of damage to plants by strong winds.

The soil stores mineral nutrients and water used by plants, as well as housing their roots. There are two general kinds of soils—mineral and the organic type called muck or peat. Mineral soils include sandy, loamy, and clayey types. Sandy and loamy soils are usually preferred for vegetable production. Soil reaction and degree of fertility can be determined by chemical analysis. The reaction of the soil determines to a great extent the availability of most plant nutrients. The degree of acid, alkaline, or neutral reaction of a soil is expressed as the pH, with a pH of 7 being neutral, points below 7 being acid, and those above 7 being alkaline. The optimum pH range for plant growth varies from one crop to another. A soil can be made more acid, or less alkaline, by applying an acid-producing chemical fertilizer such as ammonium sulfate.

The inherent fertility of soils affects production quantity, and a sound fertility program is required to maintain productivity. The ability of a soil to support plant life and produce abundant harvests is dependent on the immediately available nutrients in the soil and on the rate of release of additional nutrients that are present but not available to plants. The rate of release of these additional nutrients is affected by such factors as microbial action, soil temperature, soil moisture, and aeration. Depletion of soil fertility may occur as a result of crop removal, erosion, leaching, and volatilization, or evaporation, of nutrients.

Soil Preparation and Management

Soil preparation for vegetable growing involves many of the usual operations required for other crops. Good drainage is especially important for early vegetables because wet soil retards development. Sands are valuable in growing early vegetables because they are more readily drained than the heavier soils. Soil drainage accomplished by means of ditches or tiles is more desirable than the drainage obtained by planting crops on ridges because the former not only removes the excess water but also allows air to enter the soil. Air is essential to the growth of crop plants and to certain beneficial soil organisms making nutrients available to the plants.

When crops are grown in succession, soil rarely needs to be plowed more than once each year. Plowing incorporates sod, green-manure crops, and crop residues in the soil; destroys weeds and insects; and improves soil texture and aeration. Soils for vegetables should be fairly deep. A depth of six to eight inches (15 to 20 centimetres) is sufficient in most soils.

Soil management involves the exercise of human judgment in the application of available knowledge of crop production, soil conservation, and economics. Management should be directed toward producing the desired crops with a minimum of labour. Control of soil erosion, maintenance of soil organic matter, the adoption of crop rotation, and clean culture are considered important soil-management practices.

Soil erosion, caused by water and wind, is a problem in many vegetable-growing regions because the topsoil is usually the richest in fertility and organic matter. Soil erosion by water can be controlled by various methods. Terracing divides the land into separate drainage areas, with each area having its own waterway above the terrace. The terrace holds the water on the land, allowing it to soak into the soil and reducing or preventing gullying. In the contouring system, crops are planted in rows at the same level across the field. Cultivation proceeds along the rows rather than up and down the hill. Strip cropping consists of growing crops in narrow strips across a slope, usually on the contour. Soil erosion by wind can be controlled by the use of windbreaks of various kinds, by keeping the soil well supplied with humus, and by growing cover crops to hold the soil when the land is not occupied by other crops.

Maintenance of the organic-matter content of the soil is essential. Organic matter is a source of plant nutrients and is valuable for its effect on certain properties of the soil. Loss of organic matter is the result of the action of micro-organisms that gradually decompose it to carbon dioxide. The addition of manures and the growing of soil-improving crops are efficient means of supplying soil organic matter. Soil-improving crops are grown solely for the purpose of preparing the soil for the growth of succeeding crops. Green-manure crops, grown especially for soil improvement, are turned under while still green and usually are grown during the same season of the year as the vegetable crops. Cover crops, raised for both soil protection and improvement, are only grown during seasons when vegetable crops do not occupy the land. When a soil-improving crop is turned under, the various nutrients that have contributed to the growth of the crop are returned to the soil, adding a quantity of organic matter. Both legumes, those plants such as peas and beans having fruits and seeds formed in pods, and nonlegumes are effective soil-improving crops. The legumes, however, are more valuable, because they contribute nitrogen as well as humus. The rate of decomposition of plant material depends on the kind of crop, its stage of growth, and soil temperature and moisture. The more succulent the material is at the time it is turned under, the more quickly it decomposes. Because dry material decomposes more slowly than green material, it is desirable to turn under soil-improving crops before they are mature, unless considerable time is to elapse between the plowing and the planting of the succeeding crop. Plant material decomposes most rapidly when the soil is warm and well supplied with moisture. If soil is dry when a soil-improving crop is turned under, little or no decomposition will occur until rain or irrigation supplies the necessary moisture.

The chief benefits derived from crop rotation are the control of disease and insects and the better use of the resources of the soil. Rotation is a systematic arrangement for the growing of different crops in a more or less regular sequence on the same land. It differs from succession cropping in that rotation cropping covers a period of two, three, or more years, while in succession cropping two or more crops are grown on the same land in one year. In many regions vegetable crops are grown in rotation with other farm crops. Most vegetables grown as annual crops fit into a four-or five-year rotation plan. The system of intercropping, or companion cropping, involves the growing of two or more kinds of vegetables on the same land in the same growing season. One of the vegetables must be a small-growing and quick-maturing crop; the other must be larger and late maturing.

In the practice of clean culture, commonly followed in vegetable growing, the soil is kept free of all competing plants through frequent cultivation and the use of protective coverings, or mulches, and weed killers. In a clean vegetable field the possibility of attack by insects and disease-incitant organisms, for which plant weeds serve as hosts, is reduced.

Propagation

Propagation of crop plants, involving the formation and development of new individuals in the establishment of new plantings, is usually accomplished by the use of either seeds or the vegetative parts of plants. The first type, known as sexual propagation, is used for asparagus, bean, broccoli, cabbage, carrot, cauliflower, celery, cucumber, eggplant, leek, lettuce, lima bean, okra, onion, muskmelon, parsley, pea, pepper, pumpkin, radish, spinach, sweet corn (maize), squash, tomato, turnip, and watermelon. The second type, asexual propagation, is used for the artichoke, garlic, girasole, potato, rhubarb, and sweet potato.

Although seed cost is a small portion of the total cost of crop production, seed quality strongly affects crop success or failure. Good seed should be accurately labelled, clean, graded to size, viable, and free of diseases and insects. The reliability of the seed house is an important factor in obtaining good-quality seed. Viability, or ability to grow, and longevity, the period of viability, are characteristics of seeds of any vegetable kind. In cool, dry storage conditions, those vegetable seeds having comparatively short longevity of one to two years are okra, onion, parsley, and sweet corn. Seeds having three-year longevity are those of the asparagus, bean, carrot, leek, and pea; four-year longevity is characteristic of the beet, chard, pepper, pumpkin, and tomato seeds; longevity of five years characterizes the seeds of broccoli, cabbage, cauliflower, celery, cucumber, eggplant, lettuce, muskmelon, radish, spinach, squash, turnip, and watermelon. The dry seeds of all vegetables, when packed under vacuum in hermetically sealed cans, should remain viable for a longer period than seeds stored under less protective conditions.

Crops grown from hybrid seeds (the offspring of two or more selected parental varieties and known as F1) yield vegetables of high quantity and quality. The hybrid-seed industry is based on the production of new seed each year from the controlled pollination of selected parents found to produce the desired combination of characters in the progeny. In the early 1980s the number of F1 hybrids was increasing in Japan, the United States, and other technically advanced countries. The number of F1 hybrids varied with the kind of vegetable, but none had yet been introduced for the bean, celery, lettuce, okra, parsley, or pea.

Planting

Most vegetable crops are planted in the field where they are to grow to maturity. A few kinds are commonly started in a seedbed, established in the greenhouse or in the open, and transplanted as seedlings. Asparagus seeds are planted in a seedbed to produce crowns used for field setting. Some vegetables can be either directly seeded in the field or grown from transplants. These include broccoli, cabbage, cauliflower, celery, eggplant, leek, lettuce, onion, pepper, and tomato. The time and method of planting seeds and plants of a particular vegetable influence the success or failure of the crop. Important factors include the depth of planting, the rate of planting, and the spacing both between rows and between plants within a row.

Factors to be considered in determining the time of planting include soil and weather conditions, kind of crop, and desired harvest time. When more than one planting of a crop is made, the second and later plantings should be timed to provide a continuous harvest for the period desired. The soil temperature required for germination of the planted seed varies markedly with the various kinds of vegetables. Vegetables that will not germinate at a temperature below 60 °F (16 °C) include the bean, cucumber, eggplant, lima bean, muskmelon, okra, pepper, pumpkin, squash, and watermelon. Temperatures higher than 90 °F (32 °C) are not favourable for the germination of seeds of celery, lettuce, lima bean, parsley, pea, and spinach.

The quantity of seeds planted, or rate of planting, is mainly determined by the characteristics of the vegetable plant. The size of seeds affects the number of plants raised in a given area. Watermelon varieties, for example, differ in seed size expressed as weight. The Sugar Baby variety has an average weight of 1.4 ounces (41 grams) for 1,000 seeds; those of Blackstone variety average 4.4 ounces (125 grams). If the two are grown on two separate plots of the same area and 4.4 ounces of seeds of each cultivar are planted, the result would be three times as many of the Sugar Baby plants as the Blackstone type. Seed size and plant-growth pattern of a vegetable are major factors that govern the number of plants raised in a given area. The trend in the early 1980s was to increase plant population for many crops to achieve the greatest yield possible without impairing quality. As plant population increases per unit area, a point is reached at which each plant begins to compete for certain essential growth factors—e.g., nutrients, moisture, and light. When the population is below the level in which competition between plants occurs, increased population will have no effect on individual plant performance, and the yield per unit area will increase in direct proportion to the increment of population. When competition for essential growth factors occurs, however, yield per plant decreases.

Early harvest and economical use of space are the principal objectives of growing vegetable crops from transplants produced in a greenhouse or outdoor seedbed. It is easier to care for young plants of the cabbage, cauliflower, celery, onion, and tomato in small seedbeds than to sow the seeds in the place where the crop is to grow and mature. Land is free longer for another crop, and weeds, insects, diseases, and irrigation are more readily and economically controlled. The production of transplants is often a specialty of growers who sell their produce to other vegetable growers. The seeds may be planted at a rate three to six times that commonly used for a direct-seeded field. The young plants are removed for use as transplants when they reach the desired size and age, approximately 40 to 60 days after seeding.

Care of Crops during Growth

Practices required for a vegetable crop growing in the field include cultivation; irrigation; application of fertilizers; control of weeds, diseases, and insects; protection against frost; and the application of growth regulators if necessary.

Cultivation

Cultivation refers to stirring the soil between rows of vegetable plants. Because weed control is the most important function of cultivation, this work should be performed at the most favourable time for weed killing, when the weeds are breaking through the soil surface. When the plants are grown on ridges, it is necessary to cover the basal plant portion with soil in the case of such vegetables as asparagus, carrot, garlic, leek, onion, potato, sweet corn, and sweet potato.

Irrigation

Vegetable production requires irrigation in arid and semi-arid regions, and irrigation is frequently used as insurance against drought in more humid regions. In areas having intermittent rain for five or six months, with little or none during the remainder of the year, irrigation is essential throughout the dry season and may also be needed between rainfalls in the rainy season. The two types of land irrigation generally suited to vegetables are surface irrigation and sprinkler irrigation. A level site is required for surface irrigation, in which the water is conveyed directly over the field in open ditches at a slow, nonerosive velocity. Where water is scarce, pipelines may be used, eliminating losses caused by seepage and evaporation. The distribution of water is accomplished by various control structures, and the furrow method of surface irrigation is frequently employed because most vegetable crops are grown in rows. Sprinkler irrigation conveys water through pipes for distribution under pressure as simulated rain.

Irrigation requirements are determined by both soil and plant factors. Soil factors include texture, structure, water-holding capacity, fertility, salinity, aeration, drainage, and temperature. Plant factors include type of vegetable, density and depth of the root system, stage of growth, drought tolerance, and plant population.

Fertilizer Application

Soil fertility is the capacity of the soil to supply the nutrients necessary for good crop production, and fertilizing is the addition of nutrients to the soil. Chemical fertilizers may be used to supply the needed nitrogen, phosphorus, and potassium. Chemical tests of soil, plant, or both are used to determine fertilizer needs, and the rate of application is usually based on the fertility of the soil, the cropping system employed, the kind of vegetable to be grown, and the financial return that might be expected from the crop. Methods of fertilizer application include scattering and mixing with the soil before planting; application with a drill below the surface of the soil at the time of planting; row application before or at planting time; and row application during plant growth, also called side-dressing. Plowed down broadcast fertilizers have recently been used in combination with high analysis liquid fertilizers applied at planting or as a side-dressed band. Mechanical planting devices may employ fertilizer attachments to plant the fertilizer in the form of bands near the seed. For most vegetables, the bands are placed from two to three inches (five to 7.5 centimetres) from the seed, either at the same depth or slightly below the seed.

Weed Control

Weeds (plants growing where they are not wanted) reduce crop yield, increase production cost, and may harbour insects and diseases that attack crop plants. Methods employed to control weeds include hand weeding, mechanical cultivation, application of chemicals acting as herbicides, and a combination of mechanical and chemical means. Herbicides, selective chemical weed killers, are absorbed by the plant and induce a toxic reaction. The amount and type of herbicide that can be safely used to protect vegetable crops depends on the tolerance of the specific crops to the chemical. Most herbicides are applied as a spray, and the appropriate time for application is determined by the composition of the herbicide and the kind of vegetable crop to be treated. Preplanting treatments are applied before the crop is planted; preemergence treatments are applied after the crop is planted but before its seedlings emerge from the soil; and postemergence treatments are applied to the growing crop at a definite stage of growth.

Disease and Insect Control

The production of satisfactory crops requires rigorous disease- and insect-control measures. Crop yield may be lowered by disease or insect attack, and when plants are attacked at an early stage of growth the entire crop may be lost. Reduction in the quality of vegetable crops may also be caused by diseases and insects. Grades and standards for market vegetables usually specify strict limits on the amount of disease and insect injury that may be present on vegetables in a designated grade. Vegetables remain vulnerable to insect and disease damage after harvesting, during the marketing and handling processes. When a particular plant pest is identified, the grower can select and apply appropriate control measures. Application of insect control at the times specific insects usually appear or when the first insects are noticed is usually most effective. Effective disease control usually requires preventive procedures.

Diseases are incited by such living organisms as bacteria, fungi, and viruses. Harmful material enters the plant, develops during an incubation period, and finally causes infection, the reaction of the plant to the pathogen, or disease-producing organism. Control is possible during the inoculation and incubation phases, but when the plant reaches the infection stage it is already damaged. Typical plant diseases include mildew, leaf spots, rust, and wilt. Chemical fungicides may be used to control disease, but the use of disease-resistant plant varieties is the most effective means of control.

Vegetable breeders have developed plant varieties resistant to one or more diseases; such varieties are available for the bean, cabbage, cucumber, lettuce, muskmelon, onion, pea, pepper, potato, spinach, tomato, and watermelon.

Insects are usually controlled by the use of chemical insecticides that kill through toxic action. Many insecticides are toxic to harmful insects but do not affect bees, which are valuable for their role in pollination.

Frost Protection

Frost protection may be accomplished by increasing the amount of heat radiated from the soil when frost is likely to occur. Irrigation on the day before a predicted frost provides additional moisture in the soil to increase the amount of heat given off as infrared rays. This extra heat protects the plants from frost injury. A continuous supply of water provided by sprinkler irrigation may also protect plants from frost. As the water freezes on the plant leaves, it loses heat that is absorbed by the plant leaves, maintaining leaf temperature at 32 °F (0 °C). Because of the sugars and other substances in plant cells, the freezing point of cell sap is somewhat lower than 32 °F.

Growth Regulators

It is sometimes desirable to retard or accelerate maturity in vegetable crops. A chemical compound may be applied to prevent sprouting in onion crops. It is applied in the field sufficiently early for absorption by the still-green foliage but late enough to avoid suppressing the bulb yield. Another substance may be used to end the dormancy, or rest period, of newly harvested potato tubers intended for planting. The treated seed potatoes have uniform sprout emergence. The same substance is applied to celery from two to three weeks before harvest to elongate the stalks and increase the yield and is also used to accelerate maturity in artichokes. A chemical compound, applied when adverse weather conditions prevail during the period of fruit setting, has been used to encourage fruit set.

Harvesting

The stage of development of vegetables when harvested affects the quality of the product reaching the consumer. In some vegetables, such as the bean and pea, optimum quality is reached well in advance of full maturity and then deteriorates, although yield continues to increase. Factors determining the harvest date include the genetic constitution of the vegetable variety, the planting date, and environmental conditions during the growing season. Successive harvest dates may be obtained either by planting varieties having different maturity dates or by changing the sequence of planting dates of one particular variety. The successive method is applicable to such crops as broccoli, cabbage, cauliflower, muskmelon, onion, pea, sweet corn (maize), tomato, and watermelon. Certain varieties of the carrot, celery, cucumber, lettuce, parsley, radish, spinach, or summer squash can be sown in succession throughout most of the year in some climates, thus prolonging the harvest period.

Hand harvesting is employed along with various mechanical aids for broccoli, cabbage, cauliflower, muskmelon, and pepper crops. Many vegetables grown for processing and some vegetables destined for the fresh market are mechanically harvested. Harvesting operations may be performed by a single machine in a single step for such vegetable crops as the bean, beet, carrot, lima bean, onion, pea, potato, radish, spinach, sweet corn, sweet potato, and tomato. Designers of harvesting machinery have been working to develop a multiple-picking harvester capable of adjustment for use with more than one crop. Vegetable breeders have been able to produce vegetables with characteristics suitable for machine harvesting, including compact plant growth, uniform development, and concentrated maturity.

Storage

Fresh vegetables are living organisms, and there is a continuation of life processes in the vegetable after harvest. Changes that occur in the harvested, nonprocessed vegetable include water loss, conversion of starches to sugars, conversion of sugars to starches, flavour changes, colour changes, toughening, vitamin gain or loss, sprouting, rooting, softening, and decay.

Some changes result in quality deterioration; others improve quality in those vegetables that complete ripening after harvest. Postharvest changes are influenced by such factors as kind of crop, air temperature and circulation, oxygen and carbon dioxide contents and relative humidity of the atmosphere, and disease-incitant organisms. To maintain the fresh vegetable in the living state, it is usually necessary to slow the life processes, though avoiding death of the tissues, which produces gross deterioration and drastic differences in flavour, texture, and appearance.

Storage of vegetables contributes to price stabilization by carrying over produce from periods of high production to periods of low production. It also extends the period of consumption of many kinds of vegetables. Storage conditions can contribute to the preservation of the natural living state of the edible portion and to the prevention of deterioration through control of temperature, relative humidity, and the quality of the produce to be stored. Vegetables for storage must be free from mechanical, insect, and disease injury and should be at the proper stage of maturity.

Common (unrefrigerated) storage and cold (refrigerated) storage are the methods generally employed for vegetables. Common storage, lacking precise control of temperature and humidity, includes the use of insulated storage houses, outdoor cellars, or mounds. Cold storage allows precise regulation of temperature and humidity and maintenance of constant conditions by use of a refrigeration and ventilation system. Temporary storage, suitable only for very brief storage periods, is frequently practiced in the shipping season when large lots are accumulated for carload or truck quantities. The refrigerator car or truck is a means of temporary storage used to protect produce while it is in transit. Short-term storage may last for four or six weeks. Economic factors, such as the probability that prices will increase later in the season, encourage long-term storage of such perishable vegetables as the onion, potato, and sweet potato.

Cereals

Cereals or grains belong to the monocot families, 'Poaceae or Gramineae' and are cultivated widely to obtain the edible components of their fruit seeds. Botanically, these fruits are called 'caryopsis' and are structurally divided into endosperm, germ, and bran. Cereal grains are cultivated in huge quantities and they provide more food energy than any other type of crop, therefore, they are known as staple crops.

Cereal Production

Other than its superior nutritive value, cereals are popularly used in kitchens all over the world due to their prolific growth and plentiful production in most countries. Different countries have different cereals as their staple food; the reason for this diversity is the production statistics. For example, wheat is the most significant cereal in the diets of most European countries and India. On the other hand, rice is the primary grain used in China, Japan, South East Asia, Bangladesh, Pakistan, Brazil, Myanmar, Vietnam, and the US.

The staple food cereal in northern, central America and Africa is maize or corn, while millets and sorghum are widely consumed in India and Africa as well. Many factors like climatic differences, population choice, production quality, and others are responsible for such global differences of cereal consumption in tropical and subtropical regions. Cereals are also known as the "staff of life". However, this name is not totally justified in a true sense because you probably cannot live on cereals solely for your lifetime and retain optimum health.

Breakfast Cereal

Breakfast cereal is actually the packaged breakfast food available in almost all commercial food stores. These types of cereals are sometimes eaten cold or mixed with milk or water and then fruits are added for extra nutrition and taste. Cornflakes, porridges, and oatmeals are the best examples of breakfast cereal. Usually, these cereals are soaked or boiled to attain a softened structure, which makes them palatable. Sweeteners like honey, sugar or maple syrup are then added for a better taste. Breakfast cereals have become immensely popular in fast-moving countries, as they provide a bowl full of nutrients in a short and simple way.

Types of Cereals

The most familiar grains used for making the category of cereals include rice, maize, corn, ragi, bajra, wheat, barley, sorghum, Italian millets, and oats. Here let us discuss some of the most widely used cereal varieties.

Rice: Rice is probably the most common and popular cereal consumed all over the world; especially in tropical and temperate regions. Rice cannot grow in cold regions. It is used as the staple food in most of the countries and restaurants all over the globe prepare lip-smacking rice recipes that are truly appetizing.

Brown Rice: Brown rice is another variety of rice that is rich in the vitamin B group, particularly thiamine, riboflavin, niacin, and minerals like iron, potassium, phosphorous, and magnesium.

Maize: It is the staple cereal in continents like Africa and South America and is used as animal feed worldwide. 'Cornflakes', as we know them, are nothing but the flaked form of maize. Popcorn is also a popular corn product, which has become a favorite snack for all ages around the globe.

Wheat: This is a prime cereal consumed in temperate zones, especially in Australia, North America, Europe, and New Zealand. Wheat is a major ingredient in foods like bread, biscuits, pastries, porridge, cakes, crackers, pancakes, muesli, pies, cookies, rolls, muffins, doughnuts, gravy, and some breakfast cereals. New research suggests that eating white bread could benefit health by encouraging the growth of beneficial gut bacteria. Wheat is grown in different forms all over the globe. Some wild species of wheat are now domesticated and are grown widely in specific agricultural zones. Some such species include 'Spelt', 'Einkorn' (wheat species with one grain only), 'Emmer', and 'Durum' (species of wheat used to make semolina).

Barley: This cereal is popular and highly nutritious one. It is usually grown for malting; livestock also thrives on it in lands that are not capable of growing wheat due to financial or climatic conditions.

Sorghum: It is consumed widely in Africa and Asia and is also considered a good livestock feed.

Millet: It is grown widely in Africa and Asia. Millet porridge is extremely popular in China, Germany, and Russia. It is also used in alcoholic beverages and sometimes as bird and animal feed.

Oats: Oats were initially a staple cereal in Scotland, however now oats are very popular as breakfast cereals in almost all countries. Owing to the fiber-rich property, oats are considered vital for all ages and are used as livestock feed.

Rye: It is an important cereal grown in cold climates. It is used for making bread, beer, whiskeys, and vodka and is also used as animal fodder.

Triticale: These are manually made crops that are crossbreeds of wheat and rye. Initially, these crops were harvested only in Sweden and Scotland, but FAO recently declared that in 2009, triticale was harvested by almost 29 countries across the world.

Fonio: It is a popular crop grown in western Africa and some remote areas of India. It grows easily in semi-arid regions with poor soil quality and in a short period of only six to eight weeks. Fonio is used in bread, porridge, and beer.

Buckwheat: This cereal is popularly used in pancakes, noodles, and porridge. It is popularly used like wheat cereal, but its characteristics are not related to wheat. This cereal is high in protein and amino acids.

Quinoa: This cereal is rich in dietary fiber, iron, magnesium, and phosphorus. It is mainly grown in the Andes; however, it is also popular in North America.

Other types: Maize, rice, and wheat together make up 87% of total grain production all over the world. There are many more cereals that are used in specific regions and are not as popular as the above-mentioned ones. 'Teff' is one such cereal, grown widely in Ethiopia but hardly known elsewhere. It is rich in fiber and protein. It is often used as staple food, or sometimes used as a breakfast cereal. 'Wild rice' is grown moderately in North America; 'Amaranth' was initially a staple crop of the Aztec Empire and now it's widely cultivated in Africa.

Wheat

Wheat is the most widely cultivated cereal crop in the world, with Australia being the eighth largest producer and fourth largest exporter of wheat in the world. Wheat has come to be a firm favourite grain because of the diversity it provides in culinary applications.

Wheat Types

Many different types of wheat grain exist, with two main types being eaten in Australia, namely bread wheat (Triticum aestivum vulgare) and durum wheat (Triticum turgidum durum). The durum variety is used in the manufacture of pasta while the other type is used to produce most other wheat-based foods. Bread wheat in Australia is typically white and does not have the red colour, which typifies most bread wheat grown in the northern hemisphere.

Bread wheat is described as 'hard' or 'soft' according to its protein content. Hard wheat has more protein, including more gluten, which makes it purposeful to bake bread, while soft wheat has a much lower protein content, which when milled produces 'cake flour' for sweet biscuits and cakes.

Aside from bread wheat and durum, other types of wheat include spelt, emmer, einkorn and kamut. These wheat varieties are commonly referred to as 'ancient' grains and are increasingly being used in the manufacture of niche wheat-based food products.

- Spelt is higher in protein than common wheat and can be used in place of common wheat in most recipes. Spelt does contain gluten and should not be consumed by people who need to avoid gluten as a consequence of medically diagnosed coeliac disease.
- Emmer, otherwise known as 'farro' or 'grano farro', is staging a comeback as a gourmet ingredient, with good amounts of antioxidants.
- Einkorn is prized for having a high protein content and the highest level of lutein among wheat species.
- Kamut is an heirloom grain originating from Egypt which is higher in protein and contains more vitamin E than common wheat.

Nutrition credentials of whole grain wheat:

- Low in fat, most of which is unsaturated.

- High in carbohydrate (mainly starch) and high in insoluble dietary fibre.
- Relatively high in protein (11-13%) compared with other major grains and contains a protein complex which forms gluten.
- High in potassium and low in sodium.
- The endosperm contains glucofructan (similar in structure to inulin) which functions as a prebiotic agent and has similar properties to dietary fibre.
- Contains B-group vitamins such as thiamin, riboflavin, niacin, vitamin B6 (pyridoxine), folate and pantothenic acid.
- Contains vitamin E.
- Contains iron, zinc, magnesium, phosphorus and selenium (depending on the soil content of selenium).
- Contains small amounts of copper, manganese and calcium.
- Contains phytochemicals including lignans, phenolic acids, phytic acid, plant sterols and saponins.

Main culinary uses of wheat: Wheat is typically milled into flour which is then used to make a wide range of foods including bread, crumpets, muffins, noodles, pasta, biscuits, cakes, pastries, cereal bars, sweet and savoury snack foods, crackers, crisp-breads, sauces and confectionery (e.g. liquorice). Other culinary ap-plications of wheat include:

- Flaked, puffed and extruded wheat: All three forms are commonly used to manufacture breakfast cereals and cereal snack bars.
- Wheat bran: Added to biscuits, cakes, muffins and breads to increase the dietary fibre content. Wheat bran is also used in the manufacture of some breakfast cereals.
- Wheat germ: Can be added to breads, pastries, cakes and biscuits or sprinkled onto yoghurt, breakfast cereal or fruit dishes.
- Semolina: Mainly used for making pasta. The preferred variety of wheat for pasta is Triticum durum. It can also be cooked in milk to make semolina pudding or fried golden brown and then mixed with plenty of sugar to make Halva, as eaten in the Middle East. In Greece, semolina is used in baked cakes.
- Cous cous: Used widely in North Africa, cous cous is made from semolina grains which are sprinkled with slightly salted water and rubbed to make tiny pellets which are steamed and then dried. Instant couscous is available in Australia which needs only 5 minutes soaking in hot water.
- Burghul (also known as bulgur or cracked wheat): Is made by parboiling wheat, drying it and then coarsely grinding it. It can be steamed or boiled and used in a wide range of dishes, such as tabouleh, kofta or kibbeh.

- Kibbled wheat: Grains are cracked or broken into smaller particles and then moistened or steamed and dried. Kibbled wheat is used as an ingredient in mixed grain bread or cooked as a side dish.
- Boiled wheat: Puddings are made from boiled wheat in Lebanon and the Balkans.
- Wheat starch: Used as 'cornflour' or converted to glucose, dextrose and other sugars for use in confectionery and other manufactured foods.

Rice

Rice is an edible starchy cereal grain and the plant by which it is produced. Roughly one-half of the world population, including virtually all of East and Southeast Asia, is wholly dependent upon rice as a staple food; 95 percent of the world's rice crop is eaten by humans.

Paddy field in Minamiuonuma, Japan.

The cultivated rice plant, Oryza sativa, is an annual grass of the Gramineae family. It grows to about 1.2 metres (4 feet) in height. The leaves are long and flattened, and its panicle, or inflorescence, is made up of spikelets bearing flowers that produce the fruit, or grain.

Many cultures have evidence of early rice cultivation, including China, India, and the civilizations of Southeast Asia. However, the earliest archaeological evidence comes from central and eastern China and dates to 7000–5000 BCE. With the exception of the type called upland rice, the plant is grown on submerged land in the coastal plains, tidal deltas, and river basins of tropical, semitropical, and temperate regions. The seeds are sown in prepared beds, and when the seedlings are 25 to 50 days old, they are transplanted to a field, or paddy, that has been enclosed by levees and submerged under 5 to 10 cm (2 to 4 inches) of water, remaining submerged during the growing season.

The harvested rice kernel, known as paddy, or rough, rice, is enclosed by the hull, or husk. Milling usually removes both the hull and bran layers of the kernel, and a coating of glucose and talc is sometimes applied to give the kernel a glossy finish. Rice that is processed to remove only the husks, called brown rice, contains about 8 percent protein and small amounts of fats and is a source of thiamine, niacin, riboflavin, iron, and calcium. Rice that is milled to remove the bran as well is called white rice and is greatly diminished in nutrients. When white rice forms a major portion of the diet, there is a risk of beriberi, a disease resulting from a deficiency of thiamine and minerals.

Parboiled white rice is processed before milling to retain most of the nutrients, and enriched rice has iron and B vitamins added to it. Rice is cooked by boiling. It is eaten alone and in a great variety of soups, side dishes, and main dishes in Oriental, Middle Eastern, and many other cuisines.

The by-products of milling, including bran and rice polish (finely powdered bran and starch resulting from polishing), are used as livestock feed. Oil is processed from the bran for both food and industrial uses. Broken rice is used in brewing, distilling, and in the manufacture of starch and rice flour. Hulls are used for fuel, packing material, industrial grinding, fertilizer manufacture, and in the manufacture of an industrial chemical called furfural. The straw is used for feed, livestock bedding, roof thatching, mats, garments, packing material, and broomstraws.

Types of Rice

Rice growing in fields and paddies has three edible parts – the bran, the germ, and the endosperm – just like all other whole grains. Most of us think of "brown rice" as being synonymous with whole grain rice, but in fact whole grain rice can be many different colors, depending on the variety of rice. Most rice varieties look similarly white once they're milled to remove the bran and germ.

Rice Growing in a Field

Rice is generally grown as an annual crop, with seedlings planted or seeds sown in late spring and harvest about six months later. Flooding at various stages in the plant's life keep pests and weeds at bay. Depending on the variety and on soil and weather conditions, rice plants can grow to anywhere from 3' to 6' tall (1-2m).

Close-up of Rice Growing

Long Grain Brown Rice

Long grain rice has a long, slender kernel, four to five times longer than its width. Cooked grains are separate, light, and fluffy.

Medium Grain Brown Rice

Medium grain rice has a shorter, wider kernel (two to three times longer than its width) than long grain rice. Cooked grains are more moist and tender, and have a greater tendency to cling together than long grain.

Short Grain Brown Rice

Short grain rice has a short, plump, almost round kernel. Cooked grains are soft and cling together.

Sweet Brown Rice

Sweet rice is short and plump with a chalky white, opaque kernel. When cooked, sweet rice loses its shape and is very sticky.

Wehani Rice

This long-grain honey-red rice was naturally-bred and developed from an Indian Basmati-type seed. Like other aromatic rices, it has a distinctive nutty fragrance when cooked.

Brown Basmati Rice

India is well known for its fragrant Basmati rice, another aromatic long-grain rice with a distinct "popcorn" aroma.

Himalayan Red Rice

Also imported from India, this long-grain rice has a reddish bran layer and a nutty, complex flavor that adds visual and taste delight to any dish.

Colusari Red Rice

Grown in the Sacramento Valley of the U.S., Colusari Red Rice originated in a seed bank in Maryland. When cooked, it adds an upscale burgundy color to the plate.

Purple Thai Rice

Slightly sweeter than some other rices, Purple Thai rice was traditionally used in dessert recipes, but is now turning up in savory dishes too. Add other ingredients at the last minute, unless you want them to take on the distinctive reddish-blue hue of this rice.

Chinese Black Rice

Chinese Black Rice is a medium-grain rice with white kernels inside the black bran. Cooked, it takes on a deep purplish color.

Rice Productivity

Rice is grown in more than a hundred countries, with a total harvested area of approximately 158 million hectares, producing more than 700 million tons annually (470 million tons of milled rice). Nearly 640 million tons of rice are grown in Asia, representing 90% of global production. Sub-Saharan Africa produces about 19 million tons and Latin America some 25 million tons. In Asia and sub-Saharan Africa, almost all rice is grown on small farms of 0.5–3 ha.

Yields range from less than 1 t/ha under very poor rainfed conditions to more than 10 t/ha in intensive temperate irrigated systems. Small, and in many areas shrinking, farm sizes account for the low incomes of rice farm families. Rice grows in a wide range of environments and is productive in many situations where other crops would fail.

The highest rice yields have traditionally been obtained from plantings in high-latitude areas that have long day length and where intensive farming techniques are practiced, or in low-latitude desert areas that have very high solar energy. Southwestern Australia, Hokkaido in Japan, Spain, Italy, northern California, and the Nile Delta provide the best examples.

Global Rice Production and Consumption

Global-rice-productionMore than anywhere else in the world, rice dominates overall crop production (measured by the share of crop area harvested of rice) and overall food consumption (measured by the share of rice in total caloric intake) in rice-producing Asia.

The world's largest rice producers by far are China and India. Although its area harvested is lower than India's, China's rice production is greater due to higher yields because nearly all of China's rice area is irrigated, whereas less than half of India's rice area is irrigated. After China and India, the next largest rice producers are Indonesia, Bangladesh, Vietnam, Myanmar, and Thailand. These seven countries all had average production in 2006-08 of more than 30 million tons of paddy and together account for more than 80% of world production.

Despite Asia's dominance in rice production and consumption, rice is also very important in other parts of the world. In Africa, for example, rice has been the main staple food – defined as the food, among the three main crops, that supplies the largest amount of calories – for at least 50 years in parts of western Africa and for some countries in the Indian Ocean. In these countries, the share of calories from rice has generally not increased substantially over time. In other African countries, however, rice has displaced other staple foods because of the availability of affordable imports from Asia and rice's easier preparation, which is especially important in urban areas. On balance, in Africa, production has grown rapidly, but rice consumption has grown even faster, with the balance being met by increasing quantities of imports. Western Africa is the main producing subregion, accounting for more than 40% of African production in 2006-08. In terms of individual countries, the leading producers of paddy (2006-08) are Egypt (7.0 million t), Nigeria (3.8 million t), and Madagascar (3.2 million t).

In Latin America and the Caribbean, rice was a preferred pioneer crop in the first half of the 20th century in the savannas of Brazil, Bolivia, Colombia, Uruguay, and Venezuela, and in forest margins throughout the region. Today, rice is the most important source of calories in many Latin American countries, including Ecuador and Peru, Costa Rica and Panama, Guyana and Suriname, and the Caribbean nations of Cuba, Dominican Republic, and Haiti. It is less dominant in consumption than in Asia, however, because of the importance of wheat, maize, and beans in regional diets. Brazil is by far the largest producer, and it accounts for nearly half (46% in 2006-08) of paddy production in the region. After Brazil (11.6 million t), the largest producers are Peru and Colombia (2.5 million t each in 2006-08), followed by Ecuador (1.6 million t).

Elsewhere, the most important production centers are in the United States (California and the southern states near the Mississippi River), which produced 9.0 million t of paddy on average in 2006-08. The leading European producers are Italy, Spain, and Russia. Australia used to be an important producer, but its output has declined substantially in recent years because of recurring drought. Rice consumption in the Pacific islands has increased rapidly over the past two decades. Rice, which is all imported apart from a small amount grown in Papua New Guinea, is displacing traditional starchy root crops as a major staple due to changing tastes, ease of storage and preparation, and sometimes cost.

Yield Improvement Practices

Development of High-yielding Varieties: The Green Revolution

The 'Green Revolution' is the name given to the dramatic increase in cereal crop yields through modern agricultural inputs – irrigation, fertilizers, improved seeds, and pesticides – in the 1960s. For rice, the revolution began with the release by IRRI of the high- yielding semidwarf variety IR8 in 1966. The world average rice yield in 1960, the product of thousands of years of experience, was about 2 t/ha. Astonishingly, in only 40 more years, as the Green Revolution spread, it doubled, reaching 4 t/ha in 2000. The rice varieties and technologies developed during the Green Revolution have increased yields in some areas to 6–10 t/ha.

Widespread hunger and malnutrition, especially in Asia, made it clear that production of wheat and rice needed to increase to avoid famine.

The crisis led directly to the establishment of IRRI in 1960 and later its sister institutions in the Consultative Group on International Agricultural Research (CGIAR) system.

Fine-tuning the Technologies

Irrigation and fertilizer helped raise cereal yields, but their full impact was realized only after the development of high-yielding varieties (HYVs). These semi-dwarf rice varieties were more responsive to plant nutrients and had shorter and stiffer straw that would not fall over under the weight of heavier heads of grain. They also could mature more quickly and were insensitive to daylight length, thereby permitting more crops to be grown each year on the same land.

The first of these HYVs, named IR8, was released in 1966. Adoption of HYVs occurred quickly and about 40% of the total cereal area in Asia was planted to modern varieties by 1980. This increased to about 80% of the cropped area by 2000.

Although many of the initial HYV rice varieties dramatically raised yields, they were susceptible to pests and diseases and had cooking traits that were less appealing to consumers. Continuing investments in agricultural research led to the eventual development of second- and third-generation varieties that successfully combined high yield potential with good pest and disease resistance and preferred consumption traits.

Policy Framework

Although the Green Revolution was mainly a technology revolution, it required strong public support and policies to develop the technologies, build the required infrastructure, ensure that markets, finance, and input systems worked and that farmers had enough knowledge and economic incentive to adopt the new practices. Public interventions were especially crucial in Asia for ensuring that small farmers were not left behind, and without which the Green Revolution would have been much less pro- poor. On average, Asian countries were spending 15.4% of their total government spending on agriculture by 1972 and they doubled the real value of their agricultural expenditure by 1985.

Governments also shored up farm credit systems, subsidized key inputs – especially fertilizer, power, and water – and intervened in markets to ensure that farmers received adequate prices each year to make the technologies profitable. Many governments used their interventions to ensure that small farms did not get left behind. Substantial empirical evidence at the time showed that small farms were the more efficient producers in Asia and land reform and small farm development programs were implemented to create and support large numbers of small farms. Small farm–led agricultural growth proved to be not only more efficient but also more pro-poor, a win-win proposition for growth and poverty reduction.

Impact of the Green Revolution

The Green Revolution not only increased yields, it also reduced the production costs per kg of cereal harvested. This enabled a win-win outcome in which cereal prices could decline to the benefit of consumers even while farmers and agricultural workers increased their earnings.

However, the relationship between the Green Revolution and poverty alleviation is complex, and this has led to a large and contentious debate in the literature. Village and household studies conducted soon after the release of Green Revolution technologies raised concern that large farms were the main beneficiaries of the technologies and that poor farmers were either unaffected or made worse off.

However, some studies have found favorable longer-term impacts on inequality.

Production, Area and Yield Trends Over Time

Global rice production more than tripled between 1961 and 2010, with a compound growth rate of 2.24% per year (2.21% in rice- producing Asia). Most of the increase in rice production was due to higher yields, which increased at an annual average rate of 1.74%, compared with an annual average growth rate of 0.49% for area harvested. In absolute terms, paddy yields increased at an annual average rate of 51.1 kg/ha per year, although this rate of increase has declined in both percentage and absolute terms.

The Recent Slowdown in Rice Area and Yield Growth

Production-challengesCereal yields have continued to rise on average across Asia since the Green Revolution era, but annual growth rates are slowing. Moreover, total factor productivity has been declining, meaning that farmers now have to use higher amounts of inputs to obtain the same yields as before. Fortunately, as the figure shows, population growth in rice-producing Asia has been steadily declining for even longer. Since population growth has been the main source of rice demand growth, this trend helped to keep rice prices in check for a time. But, since the mid-1990s, population growth has exceeded rice yield growth and the gap has been growing steadily larger, creating a significant imbalance between supply and demand. This trend is evident for Asia as a whole, but also separately for East Asia, Southeast Asia, and South Asia. Stagnation in area harvested further contributed to the problem, and prices eventually began to rise. Indeed, world market rice prices rose steadily by a cumulative 67% between April 2001 and September 2007, even before the world price crisis.

There are several possible reasons for the slowdown in rice yield growth and production: displacement of cereals on better lands by more profitable crops such as groundnuts, diminishing returns to modern varieties when irrigation and fertilizer use are already high, and the fact that cereal prices have fallen relative to input costs, making additional intensification less profitable. There is also concern that pest and disease resistance to modern pesticides now slows yield growth, and that breeders have largely exploited the yield potential of major Green Revolution crops.

Environmental problems that have arisen in different areas include excessive and inappropriate use of fertilizers and pesticides that pollute waterways and kill beneficial insects and other wildlife, irrigation practices that lead to salt buildup and eventual abandonment of some of the best farming lands, increasing water scarcities in major river basins, and retreating groundwater levels in areas where more water is being pumped for irrigation than can be replenished. Some of these outcomes were inevitable as millions of largely illiterate farmers began to use modern inputs for the first time, but the problem was exacerbated by inadequate extension and training, an absence of effective regulation of water use and quality, and input pricing and subsidy policies that made modern inputs too cheap and encouraged their excessive use.

Production Challenges

Global rice consumption remains strong, driven by both population and economic growth in many Asian and African countries. But, in some developing Asian nations, including India, Vietnam, and Indonesia, per capita consumption in recent years has started to decline at a rather slow pace with rising income. In other middle- to low-income Asian countries, including the Philippines, Myanmar, Cambodia, Bangladesh, and Laos, rice consumption per capita continues to rise over time.

Despite this variation in the Asian per capita rice consumption trend, it is widely expected that per capita consumption in a majority of Asian countries will start or continue to decline in the future with rising income as people diversify their diets. But there are exceptions, such as India with its large population of ovo-vegetarians, where rice consumption patterns may not change dramatically even as income rises and a rapid increase in urbanization influences food habits.

Outside Asia, where rice is not a staple yet, per capita consumption continues to grow. This is particularly true for most countries in sub-Saharan Africa, where high population growth with changing consumer preferences is causing rapid expansion in rice consumption. In the least developed countries such as Nigeria, Tanzania, and Niger, people are moving away from tubers and cassava to rice as their income rises. Similar strong consumption growth has been evident among Middle Eastern countries with almost doubling of rice consumption in the last two decades and even more rapid increases in some Pacific island countries. Along with strong population growth, the rapid rise in per capita consumption also contributed to such rapid growth in rice demand. Rice consumption also continues to grow in Latin America and the Caribbean region with a 40% increase in the last two decades as a combination of both population growth and the steady rise in per capita consumption. Even in developed countries/regions, such as the United States and the European Union, per capita consumption continues to grow, partly because of switching from protein to more fibers in the diet and also immigration from Asian countries.

Using the population projections from the United Nations and income projections from the Food and Agricultural Policy Research Institute (FAPRI), global rice demand is estimated to rise from 439 million tons (milled rice) in 2010 to 496 million tons in 2020 and further increase to 555 million tons in 2035. This is an overall increase of 26% in the next 25 years, but the rate of growth will decline from 13% for the first 10 years to 12% in the next 15 years as population growth drops and people diversify from rice to other foods. Among the various rice-consuming regions, Asian rice consumption is projected to account for 67% of the total increase, rising from 388 million tons in 2010 to 465 million tons in 2035 despite a continuing decline in per capita consumption in China and India. In addition, 30 million tons more rice will be needed by Africa, an increase of 130% from 2010 rice consumption. In the Americas, total rice consumption is projected to rise by 33% over the next 25 years.

With further area expansion likely to be slow, global rice yields must rise faster than in the recent past if world market prices are to be stabilized at affordable levels for the billions of consumers. Globally, farmers need to produce at least 8–10 million tons more paddy rice each year—an annual increase of 1.2–1.5% over the coming decade, equivalent to an average yield increase of 0.6 t/ha during the next decade. Over the longer run, global rice consumption growth is expected to slow down but yields will have to continue to grow faster than at present because of pressure on rice lands in the developing world from urbanization, climate change, and competition from other, high-value agriculture. Rice yield growth of 1.0–1.2% annually beyond 2020 will be needed to feed the still-growing world and keep prices affordable.

Maize

Maize also known as corn, is a cereal grain first domesticated by indigenous peoples in southern Mexico about 10,000 years ago. The leafy stalk of the plant produces pollen inflorescences and separate ovuliferous inflorescences called ears that yield kernels or seeds, which are fruits.

Maize has become a staple food in many parts of the world, with the total production of maize surpassing that of wheat or rice. However, little of this maize is consumed directly by humans: most is used for corn ethanol, animal feed and other maize products, such as corn starch and corn syrup. The six major types of maize are dent corn, flint corn, pod corn, popcorn, flour corn, and sweet corn. Sugar-rich varieties called sweet corn are usually grown for human consumption as kernels, while field corn varieties are used for animal feed, various corn-based human food uses (including grinding into cornmeal or masa, pressing into corn oil, and fermentation and distillation into alcoholic beverages like bourbon whiskey), and as chemical feedstocks. Maize is also used in making ethanol and other biofuels.

Maize is widely cultivated throughout the world, and a greater weight of maize is produced each year than any other grain. In 2014, total world production was 1.04 billion tonnes. Maize is the most widely grown grain crop throughout the Americas, with 361 million metric tons grown in the United States in 2014. Approximately 40% of the crop—130 million tons—is used for corn ethanol. Genetically modified maize made up 85% of the maize planted in the United States in 2009.

Names

The word *maize* derives from the Spanish form of the indigenous Taíno word for the plant, *mahiz*. It is known by other names around the world.

The word "corn" outside North America, Australia, and New Zealand refers to any cereal crop, its meaning understood to vary geographically to refer to the local staple. In the United States, Canada, Australia, and New Zealand, *corn* primarily means maize; this usage started as a shortening of "Indian corn". "Indian corn" primarily means maize (the staple grain of indigenous Americans), but can refer more specifically to multicolored "flint corn" used for decoration.

In places outside North America, Australia, and New Zealand, *corn* often refers to maize in culinary contexts. The narrower meaning is usually indicated by some additional word, as in *sweet corn, sweetcorn, corn on the cob, baby corn,* the puffed confection known as *popcorn* and the breakfast cereal known as *corn flakes.*

Many small male flowers make up the male inflorescence, called the tassel.

In Southern Africa, maize is commonly called *mielie* (Afrikaans) or *mealie* (English), words derived from the Portuguese word for maize, *milho.*

Maize is preferred in formal, scientific, and international usage because it refers specifically to this one grain, unlike *corn*, which has a complex variety of meanings that vary by context and geographic region. *Maize* is used by agricultural bodies and research institutes such as the FAO and CSIRO. National agricultural and industry associations often include the word *maize* in their name even in English-speaking countries where the local, informal word is something other than *maize*; for example, the Maize Association of Australia, the Indian Maize Development Association, the Kenya Maize Consortium and Maize Breeders Network, the National Maize Association of Nigeria, the Zimbabwe Seed Maize Association. However, in commodities trading, *corn* consistently refers to maize and not other grains.

Structure and Physiology

The maize plant is often 3 m (10 ft) in height, though some natural strains can grow 13 m (43 ft). The stem is commonly composed of 20 internodes of 18 cm (7.1 in) length. A leaf, which grows from each node, is generally 9 cm (4 in) in width and 120 cm (4 ft) in length.

Ears develop above a few of the leaves in the midsection of the plant, between the stem and leaf sheath, elongating by around 3 millimetres (0.12 in) per day, to a length of 18 cm (7 in) with 60 cm (24 in) being the maximum alleged in the subspecies. They are female inflorescences, tightly enveloped by several layers of ear leaves commonly called husks. Certain varieties of maize have been bred to produce many additional developed ears. These are the source of the "baby corn" used as a vegetable in Asian cuisine.

The apex of the stem ends in the tassel, an inflorescence of male flowers. When the tassel is mature and conditions are suitably warm and dry, anthers on the tassel dehisce and release pollen. Maize pollen is anemophilous (dispersed by wind), and because of its large settling velocity, most pollen falls within a few meters of the tassel.

Elongated stigmas, called silks, emerge from the whorl of husk leaves at the end of the ear. They are often pale yellow and 18 cm (7 in) in length, like tufts of hair in appearance. At the end of each is a carpel, which may develop into a "kernel" if fertilized by a pollen grain. The pericarp of the fruit is fused with the seed coat referred to as "caryopsis", typical of the grasses, and the entire kernel is often referred to as the "seed". The cob is close to a multiple fruit in structure, except that the individual fruits (the kernels) never fuse into a single mass. The grains are about the size of peas, and adhere in regular rows around a white, pithy substance, which forms the ear. The maximum size of kernels is reputedly 2.5 cm (1 in). An ear commonly holds 600 kernels. They are of various colors: blackish, bluish-gray, purple, green, red, white and yellow. When ground into flour, maize yields more flour with much less bran than wheat does. It lacks the protein gluten of wheat and, therefore, makes baked goods with poor rising capability. A genetic variant that accumulates more sugar and less starch in the ear is consumed as a vegetable and is called sweet corn. Young ears can be consumed raw, with the cob and silk, but as the plant matures (usually during the summer months), the cob becomes tougher and the silk dries to inedibility. By the end of the growing season, the kernels dry out and become difficult to chew without cooking them tender first in boiling water.

Planting density affects multiple aspects of maize. Modern farming techniques in developed countries usually rely on dense planting, which produces one ear per stalk. Stands of silage maize are yet denser, and achieve a lower percentage of ears and more plant matter.

Maize is a facultative short-day plant and flowers in a certain number of growing degree days > 10 °C (50 °F) in the environment to which it is adapted. The magnitude of the influence that long nights have on the number of days that must pass before maize flowers is genetically prescribed and regulated by the phytochrome system. Photoperiodicity can be eccentric in tropical cultivars such that the long days characteristic of higher latitudes allow the plants to grow so tall that they do not have enough time to produce seed before being killed by frost. These attributes, however, may prove useful in using tropical maize for biofuels.

Immature maize shoots accumulate a powerful antibiotic substance, 2,4-dihydroxy-7-methoxy-1,4-benzoxazin-3-one (DIMBOA). DIMBOA is a member of a group of hydroxamic acids (also known as benzoxazinoids) that serve as a natural defense against a wide range of pests, including insects, pathogenic fungi and bacteria. DIMBOA is also found in related grasses, particularly wheat. A maize mutant (bx) lacking DIMBOA is highly susceptible to attack by aphids and fungi. DIMBOA is also responsible for the relative resistance of immature maize to the European corn borer (family Crambidae). As maize matures, DIMBOA levels and resistance to the corn borer decline.

Because of its shallow roots, maize is susceptible to droughts, intolerant of nutrient-deficient soils, and prone to be uprooted by severe winds.

While yellow maizes derive their color from lutein and zeaxanthin, in red-colored maizes, the kernel coloration is due to anthocyanins and phlobaphenes. These latter substances are synthesized in the flavonoids synthetic pathway from polymerisation of flavan-4-ols by the expression of maize pericarp color1 (p1) gene which encodes an R2R3 myb-like transcriptional activator of the A1 gene encoding for the dihydroflavonol 4-reductase (reducing dihydroflavonols into flavan-4-ols) while another gene (Suppressor of Pericarp Pigmentation 1 or SPP1) acts as a suppressor. The p1 gene encodes an Myb-homologous transcriptional activator of genes required for biosynthesis of red phlobaphene pigments, while the P1-wr allele specifies colorless kernel pericarp and red cobs, and unstable factor for orange1 (Ufo1) modifies P1-wr expression to confer pigmentation in kernel pericarp, as well as vegetative tissues, which normally do not accumulate significant amounts of phlobaphene pigments. The maize P gene encodes a Myb homolog that recognizes the sequence CCT/AACC, in sharp contrast with the C/TAACGG bound by vertebrate Myb proteins.

Abnormal Flowers

Maize flowers may sometimes exhibit mutations that lead to the formation of female flowers in the tassel. These mutations, *ts4* and *Ts6*, prohibit the development of the stamen while simultaneously promoting pistil development. This may cause inflorescences containing both male and female flowers, or hermaphrodite flowers.

Genetics

Exotic varieties of maize are collected to add genetic diversity when selectively breeding new domestic strains.

Maize is an annual grass in the family Gramineae, which includes such plants as wheat, rye, barley, rice, sorghum, and sugarcane. There are two major species of the genus *Zea* (out of six total): *Zea mays* (maize) and *Zea diploperennis*, which is a perennial type of teosinte. The annual teosinte variety called *Zea mays mexicana* is the closest botanical relative to maize. It still grows in the wild as an annual in Mexico and Guatemala.

Many forms of maize are used for food, sometimes classified as various subspecies related to the amount of starch each has:

- Flour corn: Zea mays var. amylacea.
- Popcorn: Zea mays var. everta.
- Dent corn : Zea mays var. indentata.
- Flint corn: Zea mays var. indurata.
- Sweet corn: Zea mays var. saccharata and Zea mays var. rugosa.
- Waxy corn: Zea mays var. ceratina.
- Amylomaize: Zea mays.

- Pod corn: Zea mays var. tunicata Larrañaga ex A. St. Hil.
- Striped maize: Zea mays var. japonica.

This system has been replaced (though not entirely displaced) over the last 60 years by multivariable classifications based on ever more data. Agronomic data were supplemented by botanical traits for a robust initial classification, then genetic, cytological, protein and DNA evidence was added. Now, the categories are forms (little used), races, racial complexes, and recently branches.

Maize is a diploid with 20 chromosomes (n=10). The combined length of the chromosomes is 1500 cM. Some of the maize chromosomes have what are known as "chromosomal knobs": highly repetitive heterochromatic domains that stain darkly. Individual knobs are polymorphic among strains of both maize and teosinte.

Barbara McClintock used these knob markers to validate her transposon theory of "jumping genes", for which she won the 1983 Nobel Prize in Physiology or Medicine. Maize is still an important model organism for genetics and developmental biology today.

The Maize Genetics Cooperation Stock Center, funded by the USDA Agricultural Research Service and located in the Department of Crop Sciences at the University of Illinois at Urbana-Champaign, is a stock center of maize mutants. The total collection has nearly 80,000 samples. The bulk of the collection consists of several hundred named genes, plus additional gene combinations and other heritable variants. There are about 1000 chromosomal aberrations (e.g., translocations and inversions) and stocks with abnormal chromosome numbers (e.g., tetraploids). Genetic data describing the maize mutant stocks as well as myriad other data about maize genetics can be accessed at MaizeGDB, the Maize Genetics and Genomics Database.

In 2005, the US National Science Foundation (NSF), Department of Agriculture (USDA) and the Department of Energy (DOE) formed a consortium to sequence the B73 maize genome. The resulting DNA sequence data was deposited immediately into GenBank, a public repository for genome-sequence data. Sequences and genome annotations have also been made available throughout the project's lifetime at the project's official site.

Primary sequencing of the maize genome was completed in 2008. On November 20, 2009, the consortium published results of its sequencing effort in *Science*. The genome, 85% of which is composed of transposons, was found to contain 32,540 genes (By comparison, the human genome contains about 2.9 billion bases and 26,000 genes). Much of the maize genome has been duplicated and reshuffled by helitrons—group of rolling circle transposons.

Breeding

Maize reproduces sexually each year. This randomly selects half the genes from a given plant to propagate to the next generation, meaning that desirable traits found in the crop (like high yield or good nutrition) can be lost in subsequent generations unless certain techniques are used.

Maize breeding in prehistory resulted in large plants producing large ears. Modern breeding began with individuals who selected highly productive varieties in their fields and then sold seed to other farmers. James L. Reid was one of the earliest and most successful developing Reid's Yellow Dent in the 1860s. These early efforts were based on mass selection. Later breeding efforts included ear to row selection, hybrids made from selected inbred lines, and the highly successful double cross hybrids using four inbred lines. University supported breeding programs were especially important in developing and introducing modern hybrids. By the 1930s, companies such as Pioneer devoted to production of hybrid maize had begun to influence long term development. Internationally important seed banks such as the International Maize and Wheat Improvement Center (CIMMYT) and the US bank at the Maize Genetics Cooperation Stock Center University of Illinois at Urbana-Champaign maintain germplasm important for future crop development.

Field of maize in Liechtenstein.

Since the 1940s the best strains of maize have been first-generation hybrids made from inbred strains that have been optimized for specific traits, such as yield, nutrition, drought, pest and disease tolerance. Both conventional cross-breeding and genetic modification have succeeded in increasing output and reducing the need for cropland, pesticides, water and fertilizer. There is conflicting evidence to support the hypothesis that maize yield potential has increased over the past few decades. This suggests that changes in yield potential are associated with leaf angle, lodging resistance, tolerance of high plant density, disease/pest tolerance, and other agronomic traits rather than increase of yield potential per individual plant.

Global Maize Program

Panorama of cornfields in Nan Province, Thailand.

CIMMYT operates a conventional breeding program to provide optimized strains. The program began in the 1980s. Hybrid seeds are distributed in Africa by the Drought Tolerant Maize for Africa project.

Genetic Modification

Genetically modified (GM) maize was one of the 26 GM crops grown commercially in 2016. Grown since 1997 in the United States and Canada, 92% of the US maize crop was genetically modified in 2016 and 33% of the worldwide maize crop was GM in 2016. As of 2011, Herbicide-tolerant maize varieties were grown in Argentina, Australia, Brazil, Canada, China, Colombia, El Salvador, the European Union, Honduras, Japan, Korea, Malaysia, Mexico, New Zealand, Philippines, the Russian Federation, Singapore, South Africa, Taiwan, Thailand, and the United States. Insect-resistant maize was grown in Argentina, Australia, Brazil, Canada, Chile, China, Colombia, Egypt, the European Union, Honduras, Japan, Korea, Malaysia, Mexico, New Zealand, Philippines, South Africa, Switzerland, Taiwan, the United States, and Uruguay.

In September 2000, up to $50 million worth of food products were recalled due to the presence of Starlink genetically modified corn, which had been approved only for animal consumption and had not been approved for human consumption, and was subsequently withdrawn from the market.

Maize is the domesticated variant of teosinte. The two plants have dissimilar appearance, maize having a single tall stalk with multiple leaves and teosinte being a short, bushy plant. The difference between the two is largely controlled by differences in just two genes, called grassy tillers-1 (*gt1*, A0A317YEZ1) and teosinte branched-1 (*tb1*, Q93WI2).

Several theories had been proposed about the specific origin of maize in Mesoamerica:

1. It is a direct domestication of a Mexican annual teosinte, Zea mays ssp. parviglumis, native to the Balsas River valley in south-eastern Mexico, with up to 12% of its genetic material obtained from Zea mays ssp. mexicana through introgression.

2. It has been derived from hybridization between a small domesticated maize (a slightly changed form of a wild maize) and a teosinte of section Luxuriantes, either Z. luxurians or Z. diploperennis.

3. It has undergone two or more domestications either of a wild maize or of a teosinte. (The term "teosinte" describes all species and subspecies in the genus Zea, excluding Zea mays ssp. mays.)

4. It has evolved from a hybridization of Z. diploperennis by Tripsacum dactyloides.

In the late 1930s, Paul Mangelsdorf suggested that domesticated maize was the result of a hybridization event between an unknown wild maize and a species of *Tripsacum*, a related genus. This theory about the origin of maize has been refuted by modern genetic testing, which refutes Mangelsdorf's model and the fourth listed above.

The teosinte origin theory was proposed by the Russian botanist Nikolai Ivanovich Vavilov in 1931 and the later American Nobel Prize-winner George Beadle in 1932. It is supported experimentally and by recent studies of the plants' genomes. Teosinte and maize are able to crossbreed and produce fertile offspring. A number of questions remain concerning the species, among them:

1. how the immense diversity of the species of sect. Zea originated.

2. how the tiny archaeological specimens of 3500–2700 BC could have been selected from a teosinte.

3. how domestication could have proceeded without leaving remains of teosinte or maize with teosintoid traits earlier than the earliest known until recently, dating from ca. 1100 BC.

The domestication of maize is of particular interest to researchers—archaeologists, geneticists, ethnobotanists, geographers, etc. The process is thought by some to have started 7,500 to 12,000 years ago. Research from the 1950s to 1970s originally focused on the hypothesis that maize domestication occurred in the highlands between the states of Oaxaca and Jalisco, because the oldest archaeological remains of maize known at the time were found there.

Connection with 'Parviglumis' Subspecies

Teosinte (top), maize-teosinte hybrid (middle), maize (bottom).

Zea mays ssp. *parviglumis*, native to the Balsas River valley in Mexico's southwestern highlands, and also known as Balsas teosinte, as being the crop wild relative teosinte genetically most similar to modern maize. This was confirmed by further studies, which refined this hypothesis somewhat. Archaeobotanical studies, published in 2009, point to the middle part of the Balsas River valley as the likely location of early domestication; this river is not very long, so these locations are not very distant. Stone milling tools with maize residue have been found in an 8,700 year old layer of deposits in a cave not far from Iguala, Guerrero.

Stucco head of the Maya maize god, 550–850 AD.

Doebley was part of the team that first published, in 2002, that maize had been domesticated only once, about 9,000 years ago, and then spread throughout the Americas.

A primitive corn was being grown in southern Mexico, Central America, and northern South America 7,000 years ago. Archaeological remains of early maize ears, found at Guila Naquitz Cave in the Oaxaca Valley, date back roughly 6,250 years; the oldest ears from caves near Tehuacan, Puebla, 5,450 B.P.

Maize pollen dated to 7,300 B.P. from San Andres, Tabasco, on the Caribbean coast has also been recovered.

As maize was introduced to new cultures, new uses were developed and new varieties selected to better serve in those preparations. Maize was the staple food, or a major staple – along with squash, Andean region potato, quinoa, beans, and amaranth – of most pre-Columbian North American, Mesoamerican, South American, and Caribbean cultures. The Mesoamerican civilization, in particular, was deeply interrelated with maize. Its traditions and rituals involved all aspects of maize cultivation – from the planting to the food preparation. Maize formed the Mesoamerican people's identity.

It is unknown what precipitated its domestication, because the edible portion of the wild variety is too small, and hard to obtain, to be eaten directly, as each kernel is enclosed in a very hard bivalve shell.

In 1939, George Beadle demonstrated that the kernels of teosinte are readily "popped" for human consumption, like modern popcorn. Some have argued it would have taken too many generations of selective breeding to produce large, compressed ears for efficient cultivation. However, studies of the hybrids readily made by intercrossing teosinte and modern maize suggest this objection is not well founded.

Spreading to the North

Around 4,500 B.P., maize began to spread to the north; it was first cultivated in what is now the United States at several sites in New Mexico and Arizona, about 4,100 B.P.

During the first millennium AD, maize cultivation spread more widely in the areas north. In particular, the large-scale adoption of maize agriculture and consumption in eastern North America took place about A.D. 900. Native Americans cleared large forest and grassland areas for the new crop.

In 2005, research by the USDA Forest Service suggested that the rise in maize cultivation 500 to 1,000 years ago in what is now the southeastern United States corresponded with a decline of freshwater mussels, which are very sensitive to environmental changes.

Cultivation

Planting

Because it is cold-intolerant, in the temperate zones maize must be planted in the spring. Its root system is generally shallow, so the plant is dependent on soil moisture. As a plant that uses C4 carbon fixation, maize is a considerably more water-efficient crop than plants that use C3 carbon fixation such as alfalfa and soybeans. Maize is most sensitive to drought at the time of silk emergence, when the flowers are ready for pollination. In the United States, a good harvest was traditionally predicted if the maize was "knee-high by the Fourth of July", although modern hybrids generally exceed this growth rate. Maize used for silage is harvested while the plant is green and the fruit immature. Sweet corn is harvested in the "milk stage", after pollination but before starch has formed, between late summer and early to mid-autumn. Field maize is left in the field until very late in the autumn to thoroughly dry the grain, and may, in fact, sometimes not be harvested until winter or even early spring. The importance of sufficient soil moisture is shown in many parts of Africa, where periodic drought regularly causes maize crop failure and consequent famine. Although it is grown mainly in wet, hot climates, it has been said to thrive in cold, hot, dry or wet conditions, meaning that it is an extremely versatile crop.

Seedlings three weeks after sowing.

Maize was planted by the Native Americans in hills, in a complex system known to some as the Three Sisters. Maize provided support for beans, and the beans provided nitrogen derived from nitrogen-fixing rhizobia bacteria which live on the roots of beans and other legumes; and squashes provided ground cover to stop weeds and inhibit evaporation by providing shade over the soil. This method was replaced by single species hill planting where each hill 60–120 cm (2.0–3.9 ft) apart was planted with three or four seeds, a method still used by home gardeners. A later technique was "checked maize", where hills were placed 40 in (1.0 m) apart in each direction, allowing cultivators to run through the field in two directions. In more arid lands, this was altered and seeds were planted in the bottom of 10–12 cm (3.9–4.7 in) deep furrows to collect water. Modern technique plants maize in rows which allows for cultivation while the plant is young, although the hill technique is still used in the maize fields of some Native American reservations. When maize is planted in rows, it also allows for planting of other crops between these rows to make more efficient use of land space.

Young stalks.

In most regions today, maize grown in residential gardens is still often planted manually with a hoe, whereas maize grown commercially is no longer planted manually but rather is planted with a planter. In North America, fields are often planted in a two-crop rotation with a nitrogen-fixing crop, often alfalfa in cooler climates and soybeans in regions with longer summers. Sometimes a third crop, winter wheat, is added to the rotation.

Mature plants showing ears.

Many of the maize varieties grown in the United States and Canada are hybrids. Often the varieties have been genetically modified to tolerate glyphosate or to provide protection against natural pests. Glyphosate is an herbicide which kills all plants except those with genetic tolerance. This genetic tolerance is very rarely found in nature.

In the midwestern United States, low-till or no-till farming techniques are usually used. In low-till, fields are covered once, maybe twice, with a tillage implement either ahead of crop planting or after the previous harvest. The fields are planted and fertilized. Weeds are controlled through the use of herbicides, and no cultivation tillage is done during the growing season. This technique reduces moisture evaporation from the soil, and thus provides more moisture for the crop. The technologies mentioned in the previous paragraph enable low-till and no-till farming. Weeds compete with the crop for moisture and nutrients, making them undesirable.

Harvesting

Mature maize ears.

Before the 20th century, all maize harvesting was by manual labour, by grazing, or by some combination of those. Whether the ears were hand-picked and the stover was grazed, or the whole plant was cut, gathered, and shocked, people and livestock did all the work. Between the 1890s and the

1970s, the technology of maize harvesting expanded greatly. Today, all such technologies, from entirely manual harvesting to entirely mechanized, are still in use to some degree, as appropriate to each farm's needs, although the thoroughly mechanized versions predominate, as they offer the lowest unit costs when scaled to large farm operations. For small farms, their unit cost can be too high, as their higher fixed cost cannot be amortized over as many units.

Harvesting maize, Jones County, Iowa.

Before World War II, most maize in North America was harvested by hand. This involved a large numbers of workers and associated social events (husking or shucking bees). From the 1890s onward, some machinery became available to partially mechanize the processes, such as one- and two-row mechanical pickers (picking the ear, leaving the stover) and corn binders, which are reaper-binders designed specifically for maize. The latter produce sheaves that can be shocked. By hand or mechanical picker, the entire ear is harvested, which then requires a separate operation of a maize sheller to remove the kernels from the ear. Whole ears of maize were often stored in corn cribs, and these whole ears are a sufficient form for some livestock feeding use. Today corn cribs with whole ears, and corn binders, are less common because most modern farms harvest the grain from the field with a combine and store it in bins. The combine with a corn head (with points and snap rolls instead of a reel) does not cut the stalk; it simply pulls the stalk down. The stalk continues downward and is crumpled into a mangled pile on the ground, where it usually is left to become organic matter for the soil. The ear of maize is too large to pass between slots in a plate as the snap rolls pull the stalk away, leaving only the ear and husk to enter the machinery. The combine separates out the husk and the cob, keeping only the kernels.

Hand-picking harvest of maize in Myanmar.

When maize is a silage crop, the entire plant is usually chopped at once with a forage harvester (chopper) and ensiled in silos or polymer wrappers. Ensiling of sheaves cut by a corn binder was formerly common in some regions but has become uncommon.

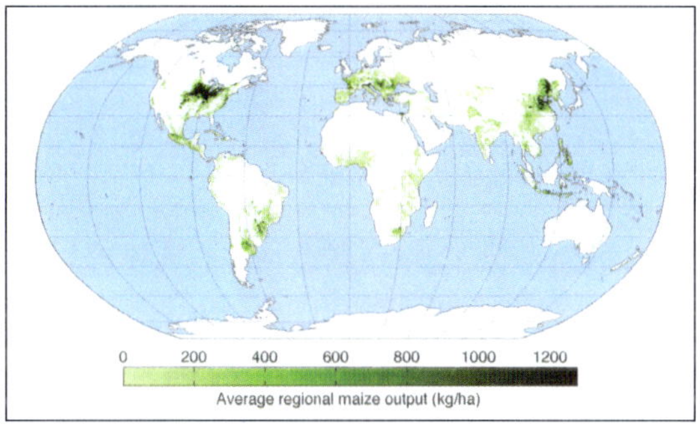

Worldwide maize production.

For storing grain in bins, the moisture of the grain must be sufficiently low to avoid spoiling. If the moisture content of the harvested grain is too high, grain dryers are used to reduce the moisture content by blowing heated air through the grain. This can require large amounts of energy in the form of combustible gases (propane or natural gas) and electricity to power the blowers.

Production

Maize is widely cultivated throughout the world, and a greater weight of maize is produced each year than any other grain. In 2014, total world production was 1.04 billion tonnes, led by the United States with 35% of the total. China produced 21% of the global total.

United States

In 2016, maize production was forecast to be over 15 billion bushels, an increase of 11% over 2014 American production. Based on conditions as of August 2016, the expected yield would be the highest ever for the United States. The area of harvested maize was forecast to be 87 million acres, an increase of 7% over 2015. Maize is especially popular in Midwestern states such as Indiana and Illinois; in the latter, it was named the state's official grain in 2017.

Pests

Insects

- African armyworm (Spodoptera exempta).
- African sugarcane borer (Eldana saccharina).
- Common armyworm (Pseudaletia unipuncta).
- Common earwig (Forficula auricularia).
- Corn delphacid (Peregrinus maidis).
- Corn leaf aphid (Rhopalosiphum maidis).
- Corn rootworms (Diabrotica spp) including Western corn rootworm (Diabrotica virgifera virgifera LeConte), Northern corn rootworm (D. barberi) and Southern corn rootworm (D. undecimpunctata howardi).
- Corn silkfly (Euxesta stigmatias).

- Asian corn borer (Ostrinia furnacalis).
- European corn borer (Ostrinia nubilalis) (ECB).
- Fall armyworm (Spodoptera frugiperda) Some sweet corn varieties have developed partial resistance to fall army worms by producing a unique 33-kD proteinase that significantly retards fall army worm growth.
- Corn earworm/Cotton bollworm (Helicoverpa zea).
- Lesser cornstalk borer (Elasmopalpus lignosellus).
- Maize weevil (Sitophilus zeamais).
- Northern armyworm, Oriental armyworm or Rice ear-cutting caterpillar (Mythimna separata).
- Southwestern corn borer (Diatraea grandiosella).
- Stalk borer (Papaipema nebris).

The susceptibility of maize to the European corn borer and corn rootworms, and the resulting large crop losses which are estimated at a billion dollars worldwide for each pest, led to the development of transgenics expressing the *Bacillus thuringiensis* toxin. “Bt maize” is widely grown in the United States and has been approved for release in Europe.

Diseases

- Rust.
- Corn smut or common smut (Ustilago maydis): a fungal disease, known in Mexico as huitlacoche, which is prized by some as a gourmet delicacy in itself.
- Northern corn leaf blight.
- Southern corn leaf blight.
- Maize downy mildew (Peronosclerospora spp.).
- Maize dwarf mosaic virus.
- Maize streak virus.
- Stewart’s wilt (Pantoea stewartii).
- Goss’s wilt (Clavibacter michiganensis).
- Grey leaf spot.
- Mal de Río Cuarto virus (MRCV).
- Stalk rot.
- Ear rot.

Uses

Human Food

Vegetable maize (sweet corn).

Maize being toasted over an open flame in India.

Maize and cornmeal (ground dried maize) constitute a staple food in many regions of the world.

Maize is central to Mexican food. Virtually every dish in Mexican cuisine uses maize. In the form of grain or cornmeal, maize is the main ingredient of tortillas, tamales, pozole, atole and all the dishes based on them, like tacos, quesadillas, chilaquiles, enchiladas, tostadas and many more. In Mexico even a fungus of maize, known as huitlacoche is considered a delicacy.

Introduced into Africa by the Portuguese in the 16th century, maize has become Africa's most important staple food crop. Maize meal is made into a thick porridge in many cultures: from the polenta of Italy, the *angu* of Brazil, the mămăligă of Romania, to cornmeal mush in the US (and hominy grits in the South) or the food called mealie pap in South Africa and *sadza*, *nshima* and *ugali* in other parts of Africa. Maize meal is also used as a replacement for wheat flour, to make cornbread and other baked products. Masa (cornmeal treated with limewater) is the main ingredient for tortillas, atole and many other dishes of Central American food.

Cut white sweet corn.

Popcorn consists of kernels of certain varieties that explode when heated, forming fluffy pieces that are eaten as a snack. Roasted dried maize ears with semihardened kernels, coated with a seasoning mixture of fried chopped spring onions with salt added to the oil, is a popular snack food in Vietnam. *Cancha*, which are roasted maize chulpe kernels, are a very popular snack food in Peru, and also appears in traditional Peruvian *ceviche*. An unleavened bread called *makki di roti* is a popular bread eaten in the Punjab region of India and Pakistan.

Chicha and *chicha morada* (purple chicha) are drinks typically made from particular types of maize. The first one is fermented and alcoholic, the second is a soft drink commonly drunk in Peru.

Dried maize *mote*, also known as hominy, is used in Mexican cuisine.

Corn flakes are a common breakfast cereal in North America and the United Kingdom, and found in many other countries all over the world.

Maize can also be prepared as hominy, in which the kernels are soaked with lye in a process called nixtamalization; or grits, which are coarsely ground hominy. These are commonly eaten in the Southeastern United States, foods handed down from Native Americans, who called the dish sagamite.

The Brazilian dessert *canjica* is made by boiling maize kernels in sweetened milk.

Maize can also be harvested and consumed in the unripe state, when the kernels are fully grown but still soft. Unripe maize must usually be cooked to become palatable; this may be done by simply boiling or roasting the whole ears and eating the kernels right off the cob. Sweet corn, a genetic variety that is high in sugars and low in starch, is usually consumed in the unripe state. Such corn on the cob is a common dish in the United States, Canada, United Kingdom, Cyprus, some parts of South America, and the Balkans, but virtually unheard of in some European countries. Corn on the cob was hawked on the streets of early 19th-century New York City by poor, barefoot "Hot Corn Girls", who were thus the precursors of hot dog carts, churro wagons, and fruit stands seen on the streets of big cities today. The cooked, unripe kernels may also be shaved off the cob and served as a vegetable in side dishes, salads, garnishes, etc. Alternatively, the raw unripe kernels may also be grated off the cobs and processed into a variety of cooked dishes, such as maize purée, tamales, *pamonhas*, *curau*, cakes, ice creams, etc.

Maize being thoroughly roasted in portable brazer in India.

Maize is a major source of starch. Cornstarch (maize flour) is a major ingredient in home cooking and in many industrialized food products. Maize is also a major source of cooking oil (corn oil) and of maize gluten. Maize starch can be hydrolyzed and enzymatically treated to produce syrups, particularly high fructose corn syrup, a sweetener; and also fermented and distilled to produce grain alcohol. Grain alcohol from maize is traditionally the source of Bourbon whiskey. Maize is sometimes used as the starch source for beer.

Within the United States, the usage of maize for human consumption constitutes about 1/40th of the amount grown in the country. In the United States and Canada, maize is mostly grown to feed livestock, as forage, silage (made by fermentation of chopped green cornstalks), or grain. Maize meal is also a significant ingredient of some commercial animal food products, such as dog food.

Boiled corn is one of the most popular street foods in Russia. Numerous streets, almost each of the parks and railway stations have a small shop that sells corncob as a snack.

Nutritional Value

Raw, yellow, sweet maize kernels are composed of 76% water, 19% carbohydrates, 3% protein, and 1% fat. In a 100-gram serving, maize kernels provide 86 calories and are a good source (10-19% of the Daily Value) of the B vitamins, thiamin, niacin, pantothenic acid (B5) and folate. In moderate amounts, they also supply dietary fiber and the essential minerals, magnesium and phosphorus whereas other nutrients are in low amounts.

Maize has suboptimal amounts of the essential amino acids tryptophan and lysine, which accounts for its lower status as a protein source. The indigenous Americans overcame this deficiency with the inclusion of beans in their diet.

Feed and Fodder for Livestock

Maize is a major source of both grain feed and fodder for livestock. It is fed to the livestock in various ways. When it is used as a grain crop, the dried kernels are used as feed. They are often kept on the cob for storage in a corn crib, or they may be shelled off for storage in a grain bin. The farm that consumes the feed may produce it, purchase it on the market, or some of both. When the grain is used for feed, the rest of the plant (the corn stover) can be used later as fodder, bedding (litter), or soil amendment. When the whole maize plant (grain plus stalks and leaves) is used for fodder, it is usually chopped all at once and ensilaged, as digestibility and palatability are higher in the ensilaged form than in the dried form. Maize silage is one of the most valuable forages for ruminants. Before the advent of widespread ensilaging, it was traditional to gather the corn into shocks after harvesting, where it dried further. With or without a subsequent move to the cover of a barn, it was then stored for weeks to several months until fed to the livestock. Today ensilaging can occur not only in siloes but also in silage wrappers. However, in the tropics maize can be harvested year-round and fed as green forage to the animals.

Chemicals

Starch from maize can also be made into plastics, fabrics, adhesives, and many other chemical products.

The corn steep liquor, a plentiful watery byproduct of maize wet milling process, is widely used in the biochemical industry and research as a culture medium to grow many kinds of microorganisms.

Chrysanthemin is found in purple corn and is used as a food coloring.

Biofuel

"Feed maize" is being used increasingly for heating; specialized corn stoves (similar to wood stoves) are available and use either feed maize or wood pellets to generate heat. Maize cobs are also used as a biomass fuel source. Maize is relatively cheap and home-heating furnaces have been developed which use maize kernels as a fuel. They feature a large hopper that feeds the uniformly sized maize kernels (or wood pellets or cherry pits) into the fire.

Maize is increasingly used as a feedstock for the production of ethanol fuel. When considering where to construct an ethanol plant, one of the site selection criteria is to ensure there is locally available feedstock. Ethanol is mixed with gasoline to decrease the amount of pollutants emitted when used to fuel motor vehicles. High fuel prices in mid-2007 led to higher demand for ethanol, which in turn led to higher prices paid to farmers for maize. This led to the 2007 harvest being one of the most profitable maize crops in modern history for farmers. Because of the relationship between fuel and maize, prices paid for the crop now tend to track the price of oil.

The price of food is affected to a certain degree by the use of maize for biofuel production. The cost of transportation, production, and marketing are a large portion (80%) of the price of food in the United States. Higher energy costs affect these costs, especially transportation. The increase in food prices the consumer has been seeing is mainly due to the higher energy cost. The effect of biofuel production on other food crop prices is indirect. Use of maize for biofuel production increases the demand, and therefore price of maize. This, in turn, results in farm acreage being diverted from other food crops to maize production. This reduces the supply of the other food crops and increases their prices.

Farm-based maize silage digester located near Neumünster in Germany, 2007. Green inflatable biogas holder is shown on top of the digester.

Maize is widely used in Germany as a feedstock for biogas plants. Here the maize is harvested, shredded then placed in silage clamps from which it is fed into the biogas plants. This process makes use of the whole plant rather than simply using the kernels as in the production of fuel ethanol.

A biomass gasification power plant in Strem near Güssing, Burgenland, Austria, began in 2005. Research is being done to make diesel out of the biogas by the Fischer Tropsch method.

Increasingly, ethanol is being used at low concentrations (10% or less) as an additive in gasoline (gasohol) for motor fuels to increase the octane rating, lower pollutants, and reduce petroleum use (what is nowadays also known as "biofuels" and has been generating an intense debate regarding the human beings' necessity of new sources of energy, on the one hand, and the need to maintain, in regions such as Latin America, the food habits and culture which has been the essence of civilizations such as the one originated in Mesoamerica; the entry, January 2008, of maize among the commercial agreements of NAFTA has increased this debate, considering the bad labor conditions of workers in the fields, and mainly the fact that NAFTA "opened the doors to the import of maize from the United States, where the farmers who grow it receive multimillion dollar subsidies and other government supports. According to OXFAM UK, after NAFTA went into effect, the price of maize in Mexico fell 70% between 1994 and 2001. The number of farm jobs dropped as well: from 8.1 million in 1993 to 6.8 million in 2002. Many of those who found themselves without work were small-scale maize growers."). However, introduction in the northern latitudes of the US of tropical maize for biofuels, and not for human or animal consumption, may potentially alleviate this.

As a result of the US federal government announcing its production target of 35 billion US gallons (130,000,000 m^3) of biofuels by 2017, ethanol production will grow to 7 billion US gallons (26,000,000 m^3) by 2010, up from 4.5 billion in 2006, boosting ethanol's share of maize demand in the US from 22.6 percent to 36.1 percent.

Commodity

Maize is bought and sold by investors and price speculators as a tradable commodity using corn futures contracts. These "futures" are traded on the Chicago Board of Trade (CBOT) under ticker symbol C. They are delivered every year in March, May, July, September, and December.

Ornamental and other uses

Some forms of the plant are occasionally grown for ornamental use in the garden. For this purpose, variegated and colored leaf forms as well as those with colorful ears are used.

Corncobs can be hollowed out and treated to make inexpensive smoking pipes, first manufactured in the United States in 1869.

Children playing in a maize kernel box.

An unusual use for maize is to create a "corn maze" (or "maize maze") as a tourist attraction. The idea of a maize maze was introduced by the American Maze Company who created a maze in Pennsylvania in 1993. Traditional mazes are most commonly grown using yew hedges, but these take several years to mature. The rapid growth of a field of maize allows a maze to be laid out using GPS at the start of a growing season and for the maize to grow tall enough to obstruct a visitor's line of sight by the start of the summer. In Canada and the US, these are popular in many farming communities.

Maize kernels can be used in place of sand in a sandboxlike enclosure for children's play.

Stigmas from female maize flowers, popularly called corn silk, are sold as herbal supplements.

Maize is used as a fish bait, called "dough balls". It is particularly popular in Europe for coarse fishing.

Additionally, feed corn is sometimes used by hunters to bait animals such as deer or wild hogs.

United States usage Breakdown

The breakdown of usage of the 12.1-billion-bushel (307-million-tonne) 2008 US maize crop was as follows, according to the World Agricultural Supply and Demand Estimates Report by the USDA.

Use	Amount		
	Million bushels	Million tonnes	Percentage
Livestock feed	5,250	133.4	43.4
Ethanol production	3,650	92.7	30.2
Exports	1,850	47.0	15.3
Production of starch, corn oil, sweeteners (hfcs, etc.)	943	24.0	7.8
Human consumption—grits, corn flour, corn meal, beverage alcohol	327	8.3	2.7

In the US since 2009/2010, maize feedstock use for ethanol production has somewhat exceeded direct use for livestock feed; maize use for fuel ethanol was 5,130 million bushels (130 million tonnes) in the 2013/2014 marketing year.

A fraction of the maize feedstock dry matter used for ethanol production is usefully recovered as DDGS (dried distillers grains with solubles). In the 2010/2011 marketing year, about 29.1 million tonnes of DDGS were fed to US livestock and poultry. Because starch utilization in fermentation for ethanol production leaves other grain constituents more concentrated in the residue, the feed value per kg of DDGS, with regard to ruminant-metabolizable energy and protein, exceeds that of the grain. Feed value for monogastric animals, such as swine and poultry, is somewhat lower than for ruminants.

Hazards

Pellagra

When maize was first introduced into farming systems other than those used by traditional native-American peoples, it was generally welcomed with enthusiasm for its productivity. However, a widespread problem of malnutrition soon arose wherever maize was introduced as a staple food. This was a mystery, since these types of malnutrition were not normally seen among the indigenous Americans, for whom maize was the principal staple food.

It was eventually discovered that the indigenous Americans had learned to soak maize in alkali-water (the process now known as nixtamalization) —made with ashes and lime (calcium oxide) since at least 1200–1500 BC by Mesoamericans and North Americans—which liberates the B-vitamin niacin, the lack of which was the underlying cause of the condition known as pellagra.

Maize was introduced into the diet of non-indigenous Americans without the necessary cultural knowledge acquired over thousands of years in the Americas. In the late 19th century, pellagra reached epidemic proportions in parts of the southern US, as medical researchers debated two theories for its origin: The deficiency theory (which was eventually shown to be true) said that pellagra was due to a deficiency of some nutrient, and the germ theory said that pellagra was caused by a germ transmitted by stable flies. A third theory, promoted by the eugenicist Charles Davenport, held that people only contracted pellagra if they were susceptible to it due to certain "constitutional, inheritable" traits of the affected individual.

Once alkali processing and dietary variety were understood and applied, pellagra disappeared in the developed world. The development of high lysine maize and the promotion of a more balanced diet have also contributed to its demise. Pellagra still exists today in food-poor areas and refugee camps where people survive on donated maize.

Allergy

Maize contains lipid transfer protein, an indigestible protein that survives cooking. This protein has been linked to a rare and understudied allergy to maize in humans. The allergic reaction can cause skin rash, swelling or itching of mucous membranes, diarrhea, vomiting, asthma and, in severe cases, anaphylaxis. It is unclear how common this allergy is in the general population.

Art

Gold maize. Moche culture 300 A.D., Larco Museum, Lima, Peru.

Water tower in Rochester, Minnesota being painted as an ear of maize.

Maize has been an essential crop in the Andes since the pre-Columbian era. The Moche culture from Northern Peru made ceramics from earth, water, and fire. This pottery was a sacred substance, formed in significant shapes and used to represent important themes. Maize was represented anthropomorphically as well as naturally.

In the United States, maize ears along with tobacco leaves are carved into the capitals of columns in the United States Capitol building. Maize itself is sometimes used for temporary architectural detailing when the intent is to celebrate the fall season, local agricultural productivity and culture. Bundles of dried maize stalks are often displayed often along with pumpkins, gourds and straw in autumnal displays outside homes and businesses. A well-known example of architectural use is the Corn Palace in Mitchell, South Dakota, which uses cobs and ears of colored maize to implement a mural design that is recycled annually. Another well known example is the *Field of Corn* sculpture in Dublin, Ohio, where hundreds of concrete ears of corn stand in a grassy field.

A maize stalk with two ripe ears is depicted on the reverse of the Croatian 1 lipa coin, minted since 1993.

Sorghum

Sorghum is also called great millet, Indian millet, milo, durra, orshallu, cereal grain plant of the grass family (Poaceae) and its edible starchy seeds. The plant likely originated in Africa, where it is a major food crop, and has numerous varieties, including grain sorghums, used for food; grass sorghums, grown for hay and fodder; and broomcorn, used in making brooms and brushes. In India sorghum is known as jowar, cholam, or jonna, in West Africa as Guinea corn, and in China as kaoliang. Sorghum is especially valued in hot and arid regions for its resistance to drought and heat.

Sorghum is a strong grass and usually grows to a height of 0.6 to 2.4 metres (2 to 8 feet), sometimes reaching as high as 4.6 metres (15 feet). Stalks and leaves are coated with a white wax, and the pith, or central portion, of the stalks of certain varieties is juicy and sweet. The leaves are about 5 cm (2 inches) broad and 76 cm (2.5 feet) long. The tiny flowers are produced in panicles that range from loose to dense; each flower cluster bears 800–3,000 kernels. The seeds vary widely among different types in colour, shape, and size, but they are smaller than those of wheat.

Broomcorn (Sorghum bicolor).

Sorghum is of a lower feed quality than corn (maize). It is high in carbohydrates, with 10 percent protein and 3.4 percent fat, and contains calcium and small amounts of iron, vitamin B1, and niacin. For human consumption, the gluten-free grain is usually ground into a meal that is made into porridge, flatbreads, and cakes. The characteristic strong flavour can be reduced by processing. The grain is also used in making edible oil, starch, dextrose (a sugar), paste, and alcoholic beverages. The stalks are used as fodder and building materials. Sweet sorghums, or sorgos, are grown mainly in the United States and southern Africa for forage and for syrup manufacture and are sometimes used in the production of ethyl alcohol for biofuel.

Poultry Processing

Poultry processing is the preparation of meat from various types of fowl for consumption by humans.

Poultry is a major source of consumable animal protein. For example, per capita consumption of poultry in the United States has more than quadrupled since the end of World War II, as the industry developed a highly efficient production system. Chickens and turkeys are the most common sources of poultry; however, other commercially available poultry meats come from ducks, geese, pigeons, quails, pheasants, ostriches, and emus.

Characteristics of Poultry

Poultry is derived from the skeletal muscles of various birds and is a good source of protein, fat, and vitamins and minerals in the diet.

Classification of Birds

Birds bred for poultry production are generally grown for a particular amount of time or until they reach a specific weight. Rock Cornish hens, narrowly defined, are a hybrid cross specifically bred to produce small roasters; in the marketplace, however, the term is used to denote a small bird, five to six weeks old, that is often served whole and stuffed. Seven-week-old chickens are classified as broilers or fryers, and those that are 14 weeks old as roasters.

Fat Content

The fat content of poultry differs in several ways from that found in red meat. Poultry has a higher proportion of unsaturated fatty acids compared with saturated fatty acids. Both turkey and chicken contain about 30 percent saturated, 43 percent monounsaturated, and 22 percent polyunsaturated fatty acids. The high levels of unsaturated fatty acids make poultry more susceptible to rancidity through the oxidation of the double bonds in the unsaturated fatty acids. Saturated fatty acids, on the other hand, do not contain double bonds in their hydrocarbon chains and are resistant to oxidation. However, this fatty acid ratio has led to the suggestion that poultry may be a more healthful alternative to red meat.

In birds fat is primarily deposited under the skin or in the abdominal cavity. Therefore, a significant amount of the fat can be removed from poultry by removing the skin before eating.

Microbial Contamination

Poultry provides an excellent medium for the growth of microorganisms. The principal spoilage bacteria found on poultry include Pseudomonas, Staphylococcus, Micrococcus, Acinetobacter, and Moraxella. In addition, poultry often supports the growth of certain pathogenic (disease-causing) bacteria, such as Salmonella.

Potential causes of contamination of poultry during the slaughtering and processing procedures include contact of the carcass with body parts that contain a high microbial load (e.g., feathers, feet, intestinal contents), use of contaminated equipment, and physical manipulation of the meat (e.g., deboning, grinding). Prevention of microbial contamination involves careful regulation and monitoring of the slaughtering and processing plants, proper handling and storage, and adequate cooking of raw and processed poultry products.

Slaughtering Procedures

Preslaughter Handling

When the birds have reached "harvest" time, they are generally taken off of feed and water. This allows their digestive tracts to empty and reduces the potential for contamination during processing.

At night the birds are caught by specially trained crews and placed into plastic or wooden transport cages. The birds are then transported to the slaughterhouse, where the trucks are often kept between sets of fans to ventilate the cages.

In the next step the birds are removed from the cages and transferred to continuously moving shackles where they are suspended by both legs. The transfer is often done in a dark room illuminated by a red light; the birds are not sensitive to the red light and this helps to keep them calm.

The handling and transfer of birds both on the farm and at the slaughterhouse can be stressful. Stress can have negative effects on the quality of the final meat product, and therefore efforts are constantly being made to improve the preslaughter processes.

Slaughtering

Stunning and Killing

After the birds have been transferred to the moving shackles, they are usually stunned by running their heads through a water bath that conducts an electric current. Stunning produces unconsciousness, but it does not kill the birds. The birds are killed either by hand or by a mechanical rotary knife that cuts the jugular veins and the carotid arteries at the neck. Any birds not killed by the machine are quickly killed by a person with a knife assigned to the bleed area. The birds are permitted to bleed for a fixed amount of time, depending on size and species (e.g., 1½ minutes for broilers). Any bird that is not properly bled will be noticeably redder after feather removal and will be condemned.

Scalding

Following bleeding, the birds go through scalding tanks. These tanks contain hot water that softens the skin so that the feathers can be removed. The temperature of the water is carefully controlled.

If retention of the yellow skin colour is desired, a soft-scald is used (about 50 °C, or 122 °F). If a white bird is desired, a higher scald temperature is used, resulting in the removal of the yellow pellicle. Turkeys and spent hens (egg-laying birds that have finished their laying cycles) are generally run at higher temperatures—59 to 60 °C (138 to 140 °F).

Defeathering

The carcasses then go through the feather-picking machines, which are equipped with rubber "fingers" specifically designed to beat off the feathers. The carcasses are moved through a sequence of machines, each optimized for removing different sets of feathers. At this point the carcasses are usually singed by passing through a flame that burns off any remaining feathers.

An extra process, called wax dipping, is often used for waterfowl, since their feathers are more difficult to remove. Following the mechanical feather picking, the carcasses are dipped in a melted, dark-coloured wax. The wax is allowed to harden and then is peeled away, pulling out the feathers at the same time. The wax is reheated and the feathers are filtered out so that the wax can be reused. This process is usually performed twice.

The blood and feathers accumulated during these early steps are generally collected and rendered to make blood meal and feather meal. The feathers from ducks and geese are often carefully collected and used for down production.

Removal of Heads and Legs

The heads of the birds go into a channel where they are pulled off mechanically; the legs of the birds are removed with a rotary knife (much like a meat slicer) either at the hock or slightly below it, depending on national custom. The carcasses drop off the shackle and are rehung by their hock onto the eviscerating shackle line. By law in the United States, the scalding and defeathering steps must be separated by a wall from the evisceration steps in order to minimize cross-contamination.

Evisceration and Inspection

At this point the preen, or oil, gland is removed from the tail and the vent is opened so that the viscera (internal organs) can be removed. Evisceration can be done either by hand (with knives) or by using complex, fully automated mechanical devices. Automated evisceration lines can operate at a rate of about 70 birds per minute. The equipment is cleaned (with relatively high levels of chlorine) after each bird.

The carcasses are generally inspected during the evisceration process. The inspection procedures in the poultry industry vary around the world and may be performed by government inspectors, veterinarians, or plant personnel, depending on a country's laws. For example, in the United States the viscera are removed and placed on the side of the bird. Inspectors from the U.S. Department of Agriculture then examine the entire bird. The plant provides each inspector with an assistant who carries out any adjustments required by the inspector (e.g., removing the entire bird or removing some part of the bird). The rejected parts are placed in a container marked "inedibles," and the contents are generally dyed (often a blue-purple), under supervision of the inspector, in order to prevent possible mixing with edible parts.

Following inspection, the carcasses are further cleaned. The viscera are separated from the carcasses, and the edible offal are removed from the inedible offal. The heart, stomach, and liver are all considered edible offal and are independently processed. Stomachs are generally cut open and the inside yellow lining of the stomach along with the stomach contents are removed.

The lungs and kidneys are removed separately from the other visceral organs using a vacuum pipe. A final inspection is often carried out at this point, and the carcasses are then washed thoroughly.

Chilling

After the carcasses have been washed, they are chilled to a temperature below 4 °C (40 °F). The two main methods for chilling poultry are water chilling and air chilling.

Water Chilling

Water chilling is used throughout North America and involves a prechilling step in which a countercurrent flow of cold water is used to lower the temperature of the carcasses. The carcasses are then moved into a chiller—a large tank specifically designed to move the carcasses through in a specific amount of time. Two tanks are used to minimize cross-contamination.

A specified overflow of water for each tank is required by law in the United States and Canada. Although this renders the chilling process very water-intensive, it helps to minimize bacterial cross-contamination by diluting the microorganisms washed off the carcasses, thereby preventing recontamination.

Water chilling leads to an increase in poultry weight, and the amount of water gained is carefully regulated. In the United States the legal limits for water pickup are 8 percent for birds going directly to market and 12 percent for birds that will be further processed (the assumption is that they will lose the extra 4 percent by the time they reach the consumer).

Air Chilling

Air chilling is the standard in Europe. The carcasses are hung by shackles and moved through coolers with rapidly moving air. The process is less energy-efficient than water chilling, and the birds lose weight because of dehydration. Air chilling prevents cross-contamination between birds. However, if a single bird contains a high number of pathogens, this pathogen count will remain on the bird. Thus, water chilling may actually result in a lower overall bacterial load, because many of the pathogens are discarded in the water.

The final temperature of the carcasses before shipment is usually about −2 to −1 °C (28 to 30 °F), just above the freezing point for poultry. In some cases a slight crusting on the surface occurs during the final chilling. For water-chilled carcasses this final chilling takes place after packaging, when the carcasses are placed in an air chiller.

Processing of Poultry

Raw Poultry Products

Whole or individual parts of birds may be packaged raw for direct sale. Poultry packaged in the United States must include instructions about safe handling, including the need to wash all equipment that has come in contact with raw poultry and the need to wash one's hands before preparing other foods. Most raw turkey is sold frozen, while most chicken is sold fresh.

Fresh Poultry

The birds are generally cut into a number of pieces, which are placed on plastic foam trays and covered with a plastic film. A "diaper" (absorbent paper with a plastic backing) is often used to catch any liquid that may be released from the meat. Fresh poultry should be used within 14 to 21 days after slaughter and generally should not be kept in the home refrigerator for more than three days. In the United States, poultry that has been frozen to a temperature of −5 to −4 °C (22 to 24 °F) and then allowed to thaw can legally be sold as "fresh."

Frozen Poultry

Most frozen poultry is vacuum-packed in plastic bags and then frozen in high-velocity freezers. The birds are kept in cold storage until needed. Before freezing, poultry may be injected with various salts, flavourings, and oils in order to increase the juiciness of the meat. Injections are usually done with a multi-needle automatic injector, and information about the added ingredients is indicated on the package label.

Frozen storage time (including poultry bought fresh and frozen in a home freezer) depends on the temperature of the freezer, the quality of the packaging, and the cycling of the freezer. For best results poultry should be used within three months. Frozen poultry products can be used directly in the frozen state or thawed first. Thawing should be done in the refrigerator or under running cold water to minimize the potential for microbial contamination.

Processed Poultry Products

Poultry may be further processed into other products. The number of processed poultry products has increased dramatically since the 1970s because of the low cost of poultry and its versatile, bland flavour.

Battering and Breading

Some poultry products are battered (e.g., with beer batter) or battered and breaded (e.g., with cracker meal, bread crumbs, or cornmeal) for frying. The meat may be either cooked or raw prior to coating. For battered and breaded poultry, the pieces are passed through a flour-based batter containing leavening and then through the breading ingredients. Many types of baked breadings have been developed to meet different tastes (e.g., Cajun or Japanese). To hold the breading to the poultry, the product is deep-fried for a short time. If the poultry is fully cooked in this process, the consumer will only have to heat the product before eating it. Chicken nuggets are a battered and breaded product that is marinated before coating.

Tumbling and Massaging

In the manufacturing of many poultry products, the meat is mixed with a variety of nonmeat ingredients, including flavourings, spices, and salt. Tumbling and massaging are gentle methods that produce a uniform meat mixture. A tumbler is a slowly rotating drum that works the meat into a smooth mixture. A massager is a large mixing chamber that contains a number of internal paddles. Cured turkey products (i.e., treated with sodium nitrite), such as turkey ham and turkey pastrami, are often tumbled or massaged during processing.

Smoking

Poultry may be smoked. Prior to smoking, the birds must be brined (soaked in a salt solution containing certain flavourings) and then allowed to dry. Smoking can be done using real wood shavings or a smoke flavouring. In the last case this must be labeled in the United States as "natural smoke flavor added."

Deboning and Grinding

Further processed poultry products leave the backs, necks, and bones available for their own processing. These materials are run through a machine called a mechanical deboner or a meat-bone separator. In general, the crushed meat and bones are continuously pressed against a screen and the edible, soft materials pushed through the screen. The resulting minced product is similar in texture to ground beef and has been used for many poultry products such as frankfurters (hot dogs) and bologna. Poultry frankfurters and bologna are made using a process similar to that for beef and pork. The meat is combined with water or ice, salt, and seasonings and chopped to emulsify the materials. The mixture is stuffed into plastic casings and cooked in a smokehouse. The meat is then quickly chilled, peeled, and vacuum-packaged. Bologna is stuffed into a larger casing and is not necessarily peeled.

References

- What-is-an-agricultural-product: thebalancesmb.com, Retrieved 21 April, 2019
- Crop, encyclopedia: nationalgeographic.org, Retrieved 15 January, 2019
- What-is-a-cash-crop, articles: worldatlas.com, Retrieved 21 July, 2019
- Cash-crop-farming, meaning, advantages, disadvantages: gardenerdy.com, Retrieved 7 May, 2019
- Van den Bosche, Peter (2005). The Law and Policy of the World Trade Organization. Cambridge, UK: Cambridge University Press. p. 42. ISBN 978-0-511-12392-4. Dumping, i.e. bringing a product onto the market of another country at a price less than the normal value of that product is condemned but not prohibited in WTO law
- Fruit, plant, reproductive-body, science: britannica.com, Retrieved 18 April, 2019
- Fruit-farming, topic: britannica.com, Retrieved 14 June, 2019
- Faodef, economic, faoinfo, waicent: fao.org, Retrieved 28 February, 2019
- Types-of-vegetables: cropsreview.com, Retrieved 19 May, 2019
- Vegetable-farming, topic: britannica.com, Retrieved 28 April, 2019
- Cereal, cereal-production, health-benefits: organicfacts.net, Retrieved 1 August, 2019
- Wheat, types-of-grains, grains: glnc.org.au, Retrieved 13 July, 2019
- Rice, plant: britannica.com, Retrieved 28 April, 2019
- Hillstrom, Kevin; Hillstrom, Laurie Collier (2003). Australia, Oceania, and Antarctica: A Continental Overview of Environmental Issues. ABC-CLIO, Inc. pp. 73–119. Retrieved April 9, 2012. ISBN 1576076954

- Wild-rice-september-grain-month, grain-month-calendar, easy-ways-enjoy-whole-grains, whole-grains: wholegrainscouncil.org, Retrieved 21 April, 2019
- Rice-productivity, rice-as-a-crop: ricepedia.org, Retrieved 17 February , 2019
- Sorghum-grain, plant: britannica.com, Retrieved 24 March , 2019
- Winkel-Shirley, B (2001). "Flavonoid biosynthesis. A colorful model for genetics, biochemistry, cell biology, and biotechnology". Plant Physiology. 126 (2): 485–93. doi:10.1104/pp.126.2.485. PMC 1540115. PMID 11402179
- Poultry-processing, technology: britannica.com, Retrieved 16 June , 2019

Chapter 3

Agricultural Processes and Resources

The various processes which are a part of agricultural production are known as agricultural processes. Some of these processes are crop choice, soil preparation, irrigation and drainage, and harvesting. This chapter discusses in detail the theories and methodologies related to these agricultural processes and resources.

Crop Choice

Crop choice is dependent upon many factors of which suitability of soil is but one. For most farmers profitability is the overruling factor, but in the case of reclaimed peatswamps there are a number of factors which influence or limit the freedom of choice and which are beyond the control of the farmer.

In the case of large peatswamps reclaimed by a Government, or government agency, the potential land use has been decided early in the scheme. Large-scale reclamation schemes need careful overall planning and the control of drainage cannot be left to the individual farmer. However, within the limits set by the controlling body it is possible to have some influence on water control at the farm level by such means as stop boards, but generally the margins are small. It is, therefore, the depth to which the groundwater is maintained in the reclaimed area that largely controls and limits the choice of crop.

In cases where an individual farmer endeavours to reclaim small stretches of swamp, he has more room for taking his own decisions. However, if his land is part of a much larger swamp he has the moral, if not legal, obligation not to disrupt farming activities on fields lying downstream. Such conditions are frequently found in long narrow interior valleys especially when under private ownership. In such cases it would be sensible to look for more participants in a concerted reclamation effort which would ensure a balanced development of the complete swamp. It would be unwise to develop one part of the same swamp for deep rooting crops demanding a drainage depth of say 90 cm, if on a neighbouring farm attempts are made to keep the water level at 40 cm for pasture development. Crop choice is also influenced by independent climatic factors and socio-economic considerations such as marketability, which are liable to fluctuations.

Peat soils are remarkably versatile in their suitability for crop growth. They have few inherent qualities which limit growth, although they require intensive and often costly improvement to natural conditions to make cropping profitable. Profitability is again largely dictated by the local economy.

Peat is a good stoneless rooting medium, it has large moisture retention capacity and hence transplanted crops establish themselves much faster than on mineral soils. Cultivations are easier than on mineral soil, even under exceptionally wet conditions. There are, however, also serious limitations to cropping:

1. Waterlogged conditions, requiring drainage.
2. Very low chemical fertility, requiring large applications of fertilizers.
3. High acidity, requiring liming.
4. Low trafficability, preventing intensive mechanized farming.

Most of these limitations can be remedied but sometimes only with capital investment. In some localities improvements are economic but in others the necessary improvements are prohibitively costly.

Most countries have already developed or should develop their own peat suitability ratings, adapted to local conditions, taking into account the limitations mentioned and based on local economic considerations. A very general guideline, based on experience in South East Asia, and which can be adopted elsewhere with similar conditions, is to limit farming to peats which are less than 2 m thick. Greater thicknesses, particularly those with a low level of management, usually have insurmountable problems. This depth requirement can be adjusted to fit local situations and where the input levels of water control, fertilization and crop protection are high deeper peats can be contemplated.

Although factors mentioned play an important role in crop choice, they are specific to a locality and should be studied and evaluated for each site.

Despite this complexity of factors, it is possible to discuss crop selection by looking at the various land use options possible under three systems of water-table management; natural conditions, somewhat improved drainage but with shallow water-tables (less than 40 cm depth), and with deep drainage (water-tables generally greater than 60 cm depth).

Choice of Crop under Natural Drainage Conditions

There is an increasing interest in developing an adapted form of agriculture which leaves the peatswamps largely in their natural state. Crop selection is directed to finding suitable swamp-adapted plant species of economic value. This type of farming would to a large extent solve the problem of subsidence. The main crops suited to such conditions are:

Sago

Sago (Metroxylon sagus, a smooth variety, and Metroxylon rumphii, the thorny variety) is grown in the natural peatswamps of Sarawak and elsewhere in the Malay-Indonesian archipelago under wild and semi-wild conditions. There is evidence that the crop could give reasonable returns, and organized production on a large scale could be lucrative. The produced starch can be used, as a food, by industry and as a raw material for the production of methanol. As a result of better vegetative dry matter production per day and a longer period of closed canopy, the sago palm is superior in potential to cassava and rice.

Raffia

The raffia palm (Raphia spp.) and papyrus (Cyperus papyrus), indigenous species in African peatswamps, can be grown commercially provided that local markets are available and the range of its industrial use can be widened.

Rice

Rice is probably the most swamp-adapted foodcrop. Wild rice (Zizania aquatica and Zizania palustris), of which a number of varieties are commercially grown in North America, even as far south as Florida, appears to offer good possibilities under the right climatic conditions in natural peatswamps. This rice, famed as a gourmet accompaniment for wild duck and game, also has a ready market as a dietary food because it is a good source of thiamine, riboflavin and nicotinic acid. Wild rice needs some regulation of water-tables for optimum growing conditions although it thrives in waterlogged conditions. Aeration of the water is however essential and regular flooding with oxygenated water is beneficial. The optimum pH is between 7.5 and 8.0 which excludes most tropical peats unless they occur near a brackish water source in the fringe areas between mangroves and freshwater swamp forest. For a number of reasons the domesticated rice (Oryza sativa) requires very good water-table control and is therefore not suitable for peatswamps. It is discussed under the section dealing with conditions of shallow drainage.

Wetland Taro

Another starch foodcrop, wetland taro (Colocasia esculenta, var. aquatilis) has many wetland cultivars. It is grown in and around the Pacific as well as in India and Africa in similar locations to wetland rice. Larger plantings require field levelling, puddling by ploughing, disking, harrowing and grading. It is also necessary to have constant water circulation to avoid footrot. Since many tropical peatswamps, particularly in their centres, have deoxygenized water, favourable conditions only occur near their edges where peat merges into mineral deposits. Nutritionally taro is a better choice of crop than rice, as the amino acid content is as high as in potatoes and the crop should be grown more widely in wetlands.

Miscellaneous Crops

Water celery (Oenanthe javanica), water spinach (Ipomoea aquatica), and the Chinese water chestnut (Eleocharis tuberosus) when grown commercially, require some expertise and the vegetable crops are limited by location and marketability. Water chestnuts, like rice, are grown in fields covered with 5-10 cm depth of water. Tubers can be eaten in salads and soups, the juice has antibiotic uses and the sedge can be made into matting although it is not very durable.

Most crops, including those mentioned above, require a plentiful supply of oxygen in the water, and relatively eutrophic conditions for good growth. The true tropical lowland peats of an oligotrophic nature are often not suitable and the best conditions are found in shallow peat and in those with an appreciable amount of mineral matter.

Choice of Crops under Improved Drainage

Pasture

By far the greatest area of reclaimed peatswamps in temperate regions is used for pasture and these only require the water-table to be lowered to about 40 cm depth. In the Netherlands a water-table depth of about 40 cm has been maintained for centuries which has minimized the rate of subsidence. The largest area of peat in the Florida Everglades is used for pasture, particularly St Augustinegrass (Stenotaphrum secundatum). Apart from requiring shallow drainage, grassland use does not necessitate intensive mechanization which is characteristic of large scale vegetable growing. Pastures can be used for direct grazing but the high water-tables lead to low surface bearing strength and pastures are unsuited to frequent and heavy traffic. In the tropics pastures on deep oligotrophic peats require high levels of fertilizer application and therefore the nature of the peat should be carefully examined and likely improvements considered in the light of local economic perspectives, before large scale pasture development is contemplated. Conditions will most likely only be favourable in a few selected localities. The growing of fodder crops such as Napier grass is probably locally more attractive than grazing. The careful selection of grass species also counteracts the effects of poaching and in Florida, St Augustinegrass is used effectively in this capacity.

Rice

Rice (Oryza sativa) has received much attention as being the crop most adapted to swamp conditions. The cultivation of rice on true tropical peats of oligotrophic nature has a number of drawbacks of which the most important are sterility, causing empty panicles, and the crop's dependence on good water control. Although the latter is only essential for obtaining high yields, the former is often the cause of complete failure. The cause of sterility is not exactly known although copper deficiency probably plays a major role and dressings of copper can alleviate the problem.

Rice fields in peatswamps where water levels are not adequately controlled may be frequently inundated by undrained rainwater and floodwaters become too high; such periods may be followed by periods of water shortage causing drought stresses in the crop. Rice growing on peat, though having some potential, therefore requires good water management involving both drainage and irrigation, and consequently level fields. The generally low or very low fertility of the majority of tropical peats, added to the problems of water control and sterility makes rice growing unprofitable in most countries, unless good management can be given and well-tried adapted cultivars are used.

Vegetables

In many countries organic soils are highly regarded for vegetable growing because of their excellent physical properties as a medium for plant growth. The choice of which vegetables to grow is dictated by many factors some of which (climate and locality) cannot be influenced by management. The minimum depth of water-table required for maximum yield and quality also limits the crop choice. Depth of water-table is often standard over large areas.

Some tropical and sub-tropical crops such as chili, soya bean and tobacco should also be mentioned here because their method of cultivation is comparable with that of vegetables. They can also be grown successfully on peat soils with shallow drainage.

Where peatswamps are in areas remote from large population centres, vegetable growing is not a commercial proposition particularly for rapidly perishable leafy vegetables. Non-perishable crops which can be stored and/or dried can often be grown profitably some distance from a market.

Horticultural Crops

In temperate regions, particularly in the Netherlands, horticultural cropping is favoured on peat soils. Here the expansion of the horticultural industry was largely based on the excellent qualities of the local peat soils. Sandy mineral soil, from nearby riverbanks or dunes, is often mixed with the surface soil further improving the inherently good physical characteristics of the peat. In Japan mineral soil dressing is also practised for wet rice cultivation. Transplanting horticultural crops is easy and successful in peat because of the good root balls formed by the plants. Horticulture can be profitable in the tropics on oligotrophic peats, drained to 50 cm depth, assuming a good local market and anadequate level of management.

Minimum water-table depth for maximum yield and quality of crops grown on organic soils.

Crop	Florida	Indiana (Depth in cm)	Minnesota
Beans	45-60	-	45
Beets (red)	-	71	-
Cabbage	45-60	66	60
Carrots	-	66	-
Celery	45	60	45
Corn	45-75	75	60
Lettuce	75-90	-	75
Mint	-	75	-
Onions	45-60	75	90
Potatoes	45-60	66	60
Parsley	35-40	-	-
Radish	35-40	-	45
Pasture-sod	30-50	45	-
Sugar cane	60-75	-	-

Choice of Crops Assuming Deep Drainage

Deep drainage is defined as greater than 60 cm depth but some crops require a depth greater than 90 cm.

Perennial Crops

The growing of perennial crops, particularly plantation crops on tropical peats, has been under study in Malaysia for some time. One of the most difficult problems to counteract is the poor root anchorage provided by the soft peat, especially for crops such as coconut palm (Cocos nucifera), oil palm (Elaeis guineensis) and rubber (Hevea brasiliensis) which become top-heavy when mature. Peat subsidence, a consequence of intensive drainage, uproots the shallow-rooting trees causing them to lean progressively and eventually topple. This problem can be partly alleviated by using dwarf varieties. Attempts are currently underway to breed a dwarf oil palm for this purpose.

It is only where the peat is shallow (less than 2 m thick) and contains an appreciable amount of mineral matter that such crops are feasible. Thus deep oligotrophic peats are unsuited because of low fertility and poor anchorage though small subsistence farmers may have marginal success when world prices are favourable.

Fruit Trees and Horticulture

The yield potential of some promising perennials and previously mentioned annuals grown on deep oligotrophic peat in South East Asia. This potential is only realized under very good management and with high inputs of fertilizers. The economic feasibility is therefore low.

Large-scale banana (Musa spp.) growing in the Ivory Coast with drainage at 80-100 cm depth on acid oligotrophic coastal peat, comparable with the lowland peats of South East Asia, indicate that yields in the first years are in the range of 25-30 t/ha but could eventually reach a level of 35-40 t/ha. However, initial investments are higher than on other soils and banana cultivation on peats is difficult and demands good management. In general this is true for most horticultural crops grown on tropical peats. Sarawak sources indicate a high potential for mulberry (Morus alba) which, however, would require the initiation of a local silk industry.

In general many fruit crops, though performing well, are plagued by bird damage, insect pests, nematode and disease problems. Therefore, these crops also need a high level of crop protection for success.

Food Crops

Root crops require well drained conditions to prevent tuber rot and large inputs of fertilizer. Under subsistence farming peat burning is traditionally practised to ensure adequate potassium levels but because this exacerbates wastage it should not be encouraged. Cassava (Manihot esculenta) or tapioca, yielding up to 50 t/ha with good management, is an important food crop on well drained deep oligotrophic peat in the tropics.

Miscellaneous Field Crops

A number of perennial or semi-perennial crops are difficult to place in the categories mentioned above. Of these pineapple (Ananas comosus) has a very good potential, because it both flourishes in the acid conditions prevalent in peat soils, and it is relatively low growing and not susceptible to being uprooted at maturity. The Malaysian pineapple industry is based predominantly on deep oligotrophic peats.

Sugar cane (Saccharum officinarum) is grown successfully on the Everglade peats in Florida and on deep coastal peats in southern states of Brazil, the latter for methanol production. Good management and assured markets are prerequisites to the economic feasibility of sugar cane growing.

Factors in Crop Selection

Prevailing Farm Conditions

An environmental scanning should first be conducted. This involves a thorough ocular inspection and other methods to obtain information on the biotics factor that can affect plant growth and yield, soil and climatic conditions prevailing in the area, and accessibility. Here the guiding rule is: know your farm first then select the right crop.

The biotic factors refer to living organisms including ruminant animals, insect and other pests, disease pathogens and weeds, as well as organisms having beneficial effects like civet cat population for the production of civet coffee and the abundance of pollinators. Where there is prevalence of a disease in a locality, susceptible crops may be excluded or a resistant variety may be selected.

The topographic features of the land like elevation, slope, terrain and aspect (the direction that a sloping land faces); and the physical and chemical properties of the soil such as texture, color, organic matter content, pH and fertility levels will determine the crops that are naturally suited. Also, the various climatic factors, such as prevailing climate type, temperature, rainfall, relative humidity, incidence of light, and frequency of typhoons will limit the choice of crops. A stable supply of water within the farm will allow wide possibilities in crop selection.

In addition, the accessibility of the farm to and from the market will influence the choice of crops. For example, cassava and oil palm should be preferably grown in farms with good roads and as close as possible to the market because the harvest is bulky and must be transported immediately due to rapid rate of degradation. Having a farm-to-market road is likewise important in stocking the farm with supplies.

The crop(s) and the variety (ies) to be grown should be selected based on their adaptability to the prevailing conditions in the farm. A useful guide is to identify the crops growing in the farm and in the neighborhood. An interview of the neighboring farmers will also provide valuable information as to the probability of success, or failure, of growing certain preferred crops.

Marketability and Profitability

For those who want to engage in cash crop farming or, at the least, ensure financial sustainability, crop selection must consider marketability and profitability. In general, this means that the crop to be selected must be high yielding. The product, be it the fruit, seed, modified root or stem, flower or foliage or any part, must have an accessible, stable and robust market. With efficient labor and use of inputs, the harvest will realize profit to finance the suceeding farm activities or generate substantial return on investment. However, market and price are dictated by many factors such as the number of competitors, supply and demand, development of new products, promotional campaign, and agribusiness cycle.

Resistance to Pests and Diseases

Regardless of the purpose of farming, it is important to be able to select a crop and variety with wide resistance to important pests and diseases. The use of susceptible varieties may result to high cost of production or, worst, total crop failure.

Available Technology

The technology for the growing of the crop must have been well established or easy to learn and apply. Likewise, certain crops are preferred because technical assistance is available locally.

Farming System

Crop selection is affected by the system of farming employed, that is, whether purely crop farming or integrated with livestock animals. Likewise, the particular crop species to be grown will depend on the crop production practices such as monoculture, multiple cropping, hedge row-strip cropping, and planting patterns.

If there is a plan to integrate crop production with free-range livestock, or where entry of astray animals is unhampered, it would be wise to install tree guards or fences. However, additional cost can be eliminated or minimized by selecting crops that are less susceptible to nibbling. Some crops, like soursop (or guyabano) and sugar apple (or atis) have anti-herbivory properties. In intercropping, it is desirable that the component crops have complementary or mutualistic relation.

Security

In the absence of security personnel or where there is no fence that will exclude intruders, crop selection may be done in favor of those which are not susceptible of thievery. This will avoid such crops as vegetables and fruit crops which can be easily harvested for food and for cash.

Availability and Cost of Planting Materials

These are important considerations especially among those who have limited capital. Take for example African oil palm. Even assuming that the projected return on investment is high and net profit can be realized in a short period of time, it would be difficult for small farmers to buy certified seedlings where the cost is about 10 times that of coconut. Plus the cost of hauling may be staggering.

Another is macapuno coconut: There are only a few tissue culture laboratories which propagate seedlings capable of producing up to 100% macapuno nuts per bunch and these laboratories may be too far away. The cost of producing seedlings being highly technical and costly, it is likewise expected that buying price would be too high. The last time that the herein author inquired, price per seedling was about 20 times more than that of normal coconut. This is one main reason why crop selection favors the normal varieties of coconut.

Availability and Cost of Labor

Where farm laborers, particularly in the community where the farm is located, are scant or cost is high, proper crop selection becomes more important. The need for farm workers should be a prime consideration because there is continuous need from land preparation to planting and up to harvest, post harvest, and marketing. For perennial crops, in fact, the work does not end with the first harvest but rather perpetually.

Soil Preparation

soil preparation is the first step to be followed when we start the agricultural process. It usually entails the loosening up of the top layer of soil i.e. top soil. Plants absorb water and nutrients from the soil through their roots, hence it is essential to prepare the soil, so we can have a healthy and bountiful produce.

The soil is prepared by the following methods:

- Ploughing,
- Levelling,
- Manuring.

Ploughing

Ploughing is the process of loosening and turning the soil. Its also known as tilling. Ploughing of soil is important for the following reasons:

- Allows roots of the plants to penetrate deeply into the soil. This firmly roots the plant.
- Loose soil also provides better aeration to the roots allowing them to breathe easily.
- It assists the growth of microbes and worms, which perform decomposing and add nutrients and humus to the soil.
- Ploughing also removes weeds and other waste materials from the field.
- It brings nutrient-rich soil to the top, which helps in the growth of plants.

The tools we use for ploughing are:

- Plough: This is the most ancient tool used for ploughing. A plough may be made of wood or a metal such as iron. It has a blade or sticks attached to the front that is used to cut through the soil. Ploughs are traditionally drawn by cattle such as ox or cows, but in modern times tractors are used. Ploughs till the soil, add manure and fertilizers and scrapes the soil.
- Hoe: A hoe is also an ancient agricultural tool. We use it to till the land, remove weeds and dig up soil. They generally have a long wooden rod with a bent iron plate at one end. The other end may be attached to an animal. Hoe was used for tilling in olden times, but these days are only used to remove weeds.
- Cultivator: This is the modern mechanism we use currently to plough our farms. It involves the use of a tractor to drive the cultivator. They dig up and pulverize the soil. Cultivators

also kill weeds and dig up unwanted vegetation. Cultivator may be more expensive than the traditional methods, but it saves a lot of time and labour force.

Levelling

Once the field is ploughed, the topsoil is quite loose. There is a strong possibility that strong winds or rain will wash away the topsoil. The soil then needs to be levelled again to ensure its strong foundation. This levelling of soil is done with an implement called the leveller, which is a heavy wooden or iron plank. Levelling of the field also helps in uniform distribution of water during irrigation. This is the final step of soil preparation.

Manuring

To increase the fertility of the soil, we add manure to the soil even before we begin sowing the seeds. We add the manure before we plough the field, so it gets properly incorporated into the soil.

Seed Selection and Sowing of Seeds

Selection of Seeds

Healthy and good quality seeds are the roots of a healthy crop. The seeds that are used to cultivate new crops have to be selected very carefully and of high quality. The good quality seeds can either be bought from different sources or farmers can produce by their own. The selection of seeds is used to improve the quality of yields. There are several diseases that are transmitted via the seeds. If the selected seeds are from the infected fields then the seed-borne diseases will cause severe problems in the agricultural process. Thus, always obtain seeds from healthy plants. Along with the diseases free and healthy seeds, farmers also need to check the germination period of the seeds, nutrients required and other benefits in terms of yield and finance. Overall, selecting good quality seeds are essential for growing strong and healthy crops.

Sowing seeds

Sowing of Seeds

Sowing seeds are necessarily the essential part of crop production. After the preparation of soil, the previously selected seeds are scattered in the field. This process is called sowing. Sowing should be done carefully and evenly. If seeds are not sown uniformly, overcrowding of crops happens. For sufficient sunlight, water and other requirements congestion need to be prevented. There are two different methods of sowing the seeds. Traditionally, sowing is done manually by hands and in some places, seed drilling machines are used.

Traditional Method or Sowing by Hands

This is the most common and generally practised method by all the farmers. This is a very simple process, in which seeds are scattered by hands. This process is called broadcasting, which is the most economical method. Here, seeds are sown either by hand or by using tools. Sowing of seeds by hand is called broadcasting. This is common, but a uniform distribution is not maintained.

Sowing by Seed Drilling Machines

This is a modern method used in some part of the world for sowing seeds. In this method, iron drills or the funnels filled with seeds are connected to a tractor. As the plough moves, seeds are distributed into the furrow made by the plough and are covered. This method is more advantageous than the traditional method. In this method, seeds are distributed evenly at regular distances and depth. This is profitable in terms of both labour and time.

Some other Methods of Sowing

There are different methods of sowing such as Broadcasting, Dibbling, Drilling, Seed dropping behind the plough, Transplanting, Hill dropping and Check row planting.

Broadcasting

Broadcasting is the process of random scattering of seed on the surface of seedbeds. It can be done manually or mechanically both. When broadcasting is done manually, uniformity of seed depends upon skill of the man. Soon after broadcasting the seeds are covered by planking or some other devices. Usually higher seed rate is obtained in this system. Mechanical broadcasters are used for large-scale work. This machine scatters the seeds on the surface of the seedbed at controlled rates.

Dibbling

Dibbling is the process of placing seeds in holes made in seedbed and covering them. In this method, seeds are placed in holes made at definite depth at fixed spacing. The equipment used for dibbling is called dibbler. It is a conical instrument used to make proper holes in the field. Small hand dibblers are made with several conical projections made in a frame. This is very time consuming process, so it is not suitable for small seeds. Mostly vegetables are sown in this way.

Seed Dropping behind the Plough

It is very common method used in villages. It is used for seed like maize, gram, peas, wheat and barley. A man drops seeds in the furrow behind the plough. Sowing behind the plough can be done by a device known as malobansa. It consists of a bamboo tube provided with a funnel shaped mouth. One man drops the seeds through the funnel and other man handles the plough and the bullocks. This is a slow and laborious method.

Transplanting

Transplanting consists of preparing seedlings in nursery and then planting these seedlings in the prepared field. It is commonly done for vegetable and flowers. It is very time consuming operation. Equipment for placing plants in the soil is called transplanter.

Hill Dropping

In this method, seeds are dropped at fixed spacing and not in a continuous stream. Thus the spacing between plant to plant in a row is constant. In case of drills, the seeds are dropped in continuous stream and the spacing between plant to plant in a row is not constant.

Check Row Planting

It is a method of planting, in which row-to-row and plant-to-plant distance is uniform. In this method, seeds are planted precisely along straight parallel furrows. The rows are always in two perpendicular directions. A machine used for check row planting is called check row planter.

Irrigation and Drainage

Irrigation and drainage is an artificial application of water to land and artificial removal of excess water from land, respectively. Some land requires irrigation or drainage before it is possible to use it for any agricultural production; other land profits from either practice to increase production. Some land, of course, does not need either. Although either practice may be, and both often are, used for nonagricultural purposes to improve the environment.

Irrigation and drainage improvements are not necessarily mutually exclusive. Often both may be required together to assure sustained, high-level production of crops.

Modern Irrigation System Planning and Construction

Water Supply

The first consideration in planning an irrigation project is developing a water supply. Water supplies may be classified as surface or subsurface. Though both surface and subsurface water come from precipitation such as rain or snow, it is far more difficult to determine the origin of subsurface water.

In planning a surface water supply, extensive studies must be made of the flow in the stream or river that will be used. If the streamflow has been measured regularly over a long period, including times of drought and flood, the studies are greatly simplified. From streamflow data, determinations can be made of the minimum, maximum, average daily, and average monthly flows; the size of dams, spillways, and downstream channel; and the seasonal and carry-over storage needed. If adequate streamflow data are not available, the streamflow may be estimated from rain and snow data, or from flow data from nearby streams that have similar climatic and physiographic conditions.

The quality, as well as the quantity, of surface water is a factor. The two most important considerations are the amount of silt carried and the kind and amount of salts dissolved in the water. If the silt content is high, sediment will be deposited in the reservoir, increasing maintenance costs and decreasing useful life periods. If the salt concentration is high, it may damage crops or accumulate in the soil and eventually render it unproductive.

Subsurface sources of water must be as carefully investigated as surface sources. In general, less is known about subsurface supplies of water than about surface supplies, so, therefore, subsurface supplies are harder to investigate. Engineers planning a project need to know the extent of the basic geological source of water (the aquifer), as well as the amount the water level is lowered by pumping and the rate of recharge of the aquifer. Often the only way for the engineer to obtain these data reliably is to drill test wells and make on-site measurements. Ideally, a project is planned so as not to use more subsurface water than is recharged. Otherwise, the water is said to be "mined," meaning that it is being used up as a natural resource and its use is considered unsustainable.

Two sources of water not often thought of by the layperson are dew and sewage or wastewater. In certain parts of the world, Israel and part of Australia, for example, where atmospheric conditions are right, sufficient dew may be trapped at night to provide water for irrigation. Elsewhere the supply of wastewater from some industries and municipalities is sufficient to irrigate relatively small acreages. Recently, due to greater emphasis on purer water in streams, there has been increased interest in this latter practice.

In some countries (Egypt for example) sewage is a valuable source of water. In others, such as the United States, irrigation is looked upon as a means of disposing of sewer water as a final step in the wastewater treatment process. Unless the water contains unusual chemical salts, such as sodium, it is generally of satisfactory quality for agricultural irrigation. Where the practice is used primarily as a means of disposal, large areas are involved and the choice of crop is critical. Usually only grass or trees can withstand the year-round applications.

Before a water supply can be assured, the right to it must be determined. Countries and states have widely varying laws and customs that determine ownership of water. If the development of a water supply is for a single purpose, then the determination of ownership may be relatively simple; but if the development is multipurpose, as most modern developments are, ownership may be difficult to determine, and agreements must be worked out among countries, states, municipalities, and private owners.

The area that can be irrigated by a water supply depends on the weather, the type of crop grown, and the soil. Numerous methods have been developed to evaluate these factors and predict average annual volume of rainfall needed. Some representative annual amounts of rainfall needed for cropland in the western United States are 305 to 760 mm (12 to 30 inches) for cereal grains and 610 to 1,525 mm (24 to 60 inches) for forage. In the Near East, cotton needs about 915 mm (36 inches), whereas rice may require two to three times that amount. In humid regions of the United States, where irrigation supplements rainfall, grain crops may require 150 to 230 mm (6 to 9 inches) of water. In addition to satisfying the needs of the crop, allowances must be made for water lost directly to evaporation and during transport to the fields.

Transport Systems

The type of transport system used for an irrigation project is often determined by the source of the water supply. If a surface water supply is used, a large canal or pipeline system is usually required to carry the water to the farms because the reservoir is likely to be distant from the point of use. If subsurface water drawn from wells is used, a much smaller transport system is needed, though canals or pipelines may be used. The transport system will depend as far as possible on gravity flow, supplemented if necessary by pumping. From the mains, water flows into branches, or laterals, and finally to distributors that serve groups of farms. Many auxiliary structures are required, including weirs (flow-diversion dams), sluices, and other types of dams. Canals are normally lined with concrete to prevent seepage losses, control weed growth, eliminate erosion hazards, and reduce maintenance. The most common type of concrete canal construction is by slip forming. In this type of construction, the canal is excavated to the exact cross section desired and the concrete placed on the earth sides and bottom.

Pipelines may be constructed of many types of material. The larger lines are usually concrete whereas laterals may be concrete, cement–asbestos, rigid plastic, aluminum, or steel. Although pipelines are more costly than open conduits, they do not require land after construction, suffer little evaporation loss, and are not troubled by algae growth.

Water Application

After water reaches the farm it may be applied by surface, subsurface, or sprinkler-irrigation methods. Surface irrigation is normally used only where the land has been graded so that uniform slopes exist. Land grading is not necessary for other methods. Each method includes several variations, only the more common of which are considered here.

Irrigation sprinklers.

Surface irrigation systems are usually classed as either flood or furrow systems. In the flood system, water is applied at the edge of a field and allowed to move over the entire surface to the opposite side of the field. Grain and forage crops are quite often irrigated by flood techniques. The furrow system is used for row crops such as corn (maize), cotton, sugar beets, and potatoes. Furrows are plowed between crop rows and the water is run in the furrows. In either type of surface irrigation systems, waste-water ditches at the lower edge of the fields permit excess water to be removed for use elsewhere and to prevent waterlogging.

Subirrigation is a less common method. An impermeable layer must be located below, but near, the root zone of the crop so that water is trapped in the root zone. If this condition exists, water is applied to the soil through tile drains or ditches.

Sprinklers have been used increasingly to irrigate agricultural land. Little or no preparation is needed, application rates can be controlled, and the system may be used for frost protection and the application of chemicals, such as pesticides, herbicides, and fertilizers. Sprinklers range from those that apply water in the form of a mist to those that apply an inch or more per hour.

Evaporation and Seepage Control

Various techniques have been tried to reduce losses of irrigation water. Two major sources of loss, particularly from surface supplies and surface systems, are evaporation and seepage from reservoirs and canals. Many studies have been made of techniques to suppress evaporation. One of the more promising appears to be application of a special alcohol film on the surface, which retards evaporation by about 30 percent and does not reduce the quality of the water. The primary problem in its use is that it is fragile; a strong wind can blow it apart and expose the water to evaporation.

Seepage has largely been controlled by lining main and distribution channels with impervious material, typically concrete. Other materials used are asphalt and plastic film, though plastic tends to deteriorate if it is exposed to sunlight.

Typical Systems

The typical surface irrigation system utilizes a publicly developed water supply—e.g., a river-basin reservoir. The public project also constructs the main canals to take water from the reservoir to the agricultural land. In general the canals flow by gravity, but lift stations are often required. Supply and field canals are used to bring the water to the individual field, where it is applied to the land either by furrow or by flooding method.

Until recently most sprinkler-irrigation systems depended on privately developed water supplies, but many modern sprinkler systems have been able to draw on public water supplies. In either case, a pump is required to pump water from a large well (3,785 litres, or 1,000 gallons, per minute or more) or a supply canal. The water goes into the system main and thence to a sprinkler unit. Many automatic or semiautomatic moving sprinkler systems travel over the field applying water. Two common units are the so-called centre pivot and the travelling sprinkler. The centre-pivot unit is anchored at the centre of the field; a long lateral (arm) with sprinklers mounted on it sweeps the field in a circle. The system has the disadvantage of missing the corners of a square field. A travelling sprinkler is mounted on a trailer and propelled across the field in a lane that has been left unplanted. The unit drags a flexible hose connected to the main supply line. When it reaches the end of the lane, it is automatically shut off and can be moved to the next lane. Despite some shortcomings, all sprinkler systems are effective in applying a controlled amount of water at a high level of efficiency with a minimum of labour.

Modern Drainage System Planning and Construction

Planning a System

The planning and design of drainage systems is not an exact science. Although there have been many advances in soil and crop science, techniques have not been developed for combining the basic principles involved into precise designs. One of the primary reasons for difficulty in applying known theory is the capricious variability of natural soil in contrast to the idealized soils required to develop a theory.

The type of drainage system designed depends on many factors, but the most important is the type of soil, which determines whether water will move through rapidly enough to use subsurface drainage. Soils that have a high percentage of sand- and silt-size particles and a low percentage of clay-size particles usually will transmit water rapidly enough to make subsurface drainage feasible. Soils that are high in clay-size particles usually cannot be drained by subsurface improvements. It is essential to consider soil properties to a depth of 1.5 to 1.8 metres (5 to 6 feet) because the layer in the soil that transmits water the slowest controls the design, and subsurface improvements may be installed to these depths.

The topography or slope of the land is also important. In many cases, land in need of drainage is so flat that a contour map showing elevations 30 cm (12 inches) or 15 cm (6 inches) apart is used to identify trouble spots and possible outlets for drainage water. Often an outlet can be developed only by collective community action. The rainfall patterns, the crops to be grown, and the normal height of the water table also are considered. If heavy rainfall is not probable during critical stages of crop growth, less extensive drainage improvements may suffice. The capacity of the system is governed in part by the growth pattern of the crop, its planting date, critical stages of growth, tolerance of excess water, harvest date, and value.

In some areas the normal water level in the soil is high, in others low; this variable is always investigated before a drainage system is planned.

Types of Drainage Systems

Drainage systems may be divided into two categories, surface and subsurface. Each has several components with similar functions but different names. At the lower, or disposal, end of either system is an outlet. In order of decreasing size, the components of a surface system are the main collection ditch, field ditch, and field drain; and for a subsurface system, main, submain, and lateral conduits from the submain. The outlet is the point of disposal of water from the system; the main carries water to the outlet; the submain or field ditch collects water from a number of smaller units and carries it to the main; and the lateral or field drain, the smallest unit of the system, removes the water from the soil.

The outlet for a drainage system may be a natural stream or river or a large constructed ditch. A constructed ditch usually is trapezoidal in section with side banks flat enough to be stable. Grass may be grown on the banks, which are kept clear of trees and brush that would interfere with the flow of water.

A surface drainage system removes water from the surface of the soil and to approximately the bottom of the field ditches. A surface system is the only means for drainage improvement on soils that transmit water slowly. Individual surface drains also are used to supplement subsurface systems by removing water from ponded areas.

The field drains of a surface system may be arranged in many patterns. Probably the two most widely used are parallel drains and random drains. Parallel drains are channels running parallel to one another at a uniform spacing of a few to several hundred metres apart, depending on the soil and the slope of the land. Random drains are channels that run to any low areas in the field. The parallel system provides uniform drainage, whereas the random system drains only the low areas connected by channels. In either case the channels are shallow with flat sides and may be farmed like the rest of the field. Crops are usually planted perpendicular to the channels so that the water flows between the rows to the channels.

Some land grading of the fields where surface drains are installed is usually essential for satisfactory functioning. Land grading is the shaping of the field so that the land slopes toward the drainage channels. The slope may be uniform over the entire field or it may vary from part to part. Historically, the calculations necessary for planning land grading were time-consuming, a factor that restricted the alternatives available for final design. Today, computer models rapidly explore many possibilities before a final land grading design is selected.

In a subsurface drainage system, often called a tile system, all parts except the outlet are located below the surface of the ground. It provides better drainage than a surface system because it removes water from the soil to the depth of the drain, providing plants a greater mass of soil for root development, permitting the soil to warm up faster in the spring, and maintaining a better balance of bacterial action, the air in the soil, and other factors needed for maximum crop growth.

The smallest component of the subsurface system, the lateral, primarily removes water from the soil. The laterals may be arranged in either a uniform or a random pattern. The choice is governed by the crop grown and its value, the characteristics of the soil, and the precipitation pattern.

The primary decision required for a system with uniform laterals is their depth and spacing. In general, the deeper the laterals can be emplaced, the farther apart they can be spaced for an equivalent degree of drainage. Laterals usually are spaced from 24 to 91 metres (80 to 300 feet) apart and 0.9 to 1.5 metres (3 to 5 feet) deep.

Subsurface drainage systems are as important in many irrigated areas as they are in humid areas. A drainage system is needed on irrigated lands to control the water table and ensure that water will be able to move through a soil, thus keeping salts from accumulating in the root zone and making the soil unproductive.

Construction and Maintenance

Most subsurface drains are constructed by excavating a trench, installing a tile, and backfilling the trench. Control of the machines to assure proper slope of the drain has been a major problem, but recent development in excavation technology, including the use of laser beams for grade control, have helped to solve it. Traditionally, clay or concrete tile has been the principal material used, but many types of perforated plastic tubes are now employed. An advantage is the reduction in weight of the material handled.

With proper maintenance, drainage systems give relatively long life. Selected herbicides are often applied to keep woody growth and water weeds out of the channels. Grates are usually installed over outlets to prevent rodents and burrowing animals from building nests.

Surface drainage systems need almost yearly maintenance to assure the slope and cross section of the channels and the slope of the graded areas because the slopes are so flat that small changes in the ground surface can make marked changes in the ability of a system to function.

Subsurface systems need periodic inspection but usually require little servicing. The outlet of the system and infrequent structural failure of the material are the usual points for service.

Land Reclamation through Irrigation and Drainage

The continual need for increased food and fibre production requires the continued development of new agricultural lands and increased efficiency of existing agricultural areas. Development of new agricultural areas is rarely possible without irrigation or drainage systems or both. Easily recognized improvements are the large-scale river-basin projects designed for flood control, irrigation, and power generation. Such projects are in various stages of design or construction in many countries of the world—for example, China, India, Egypt, Iran, Australia, and the United States. In almost all cases drainage of the irrigated lands is considered a companion requirement. If possible, the drainage improvements are subsurface.

A combination of drainage and irrigation is used to reclaim large areas of land that have been abandoned because of salt accumulation. In this case subsurface drainage systems must be installed so that high water tables are lowered and pure water flushed through the soil, dissolving the salts and carrying them away in the drainage water. Large areas in the United States, India, and the Middle East are potentially available for reclamation by this technique.

The people of the Netherlands have reclaimed land from the sea by the use of drainage. Since the IJsselmeer (formerly Zuiderzee) barrier dam was closed in 1932, converting this large body of water into a freshwater lake, the Dutch have been continually enclosing and reclaiming smaller bodies (polders). After dikes are built around a polder, the area is drained by pumping out the water. Drainage channels and, in many places, subsurface drains are installed so that the root zone of crops can be drained. After this, cropping is started as the last step in the reclamation process.

The development of land-clearing machinery and surface-drainage techniques has made it possible to clear and drain tropical lands for agricultural production. The first step is the removal of trees, brush, and other tropical growth. Outlet ditches are constructed, followed by drains. In some cases subsurface drains are possible, but more often the soils and rainfall conditions combine to make this improvement impractical. Surface drains are installed on a uniform pattern and the land is smoothed or graded. Drainage systems on newly reclaimed tropical land require special attention while the soils are stabilizing, and some reconstruction is often needed after the soil stabilization is complete.

Irrigation and Drainage throughout the World

The Food and Agriculture Organization of the United Nations (FAO) keeps the most complete statistics on irrigated lands; it estimates that in the entire world some 275 million hectares (680 million acres) are irrigated. More than 130 countries report some acreage under irrigation. Asia irrigates close to 70 percent of the total area of the world that is irrigated; most of this is the large surface-irrigated rice-producing areas of China, India, Pakistan, and Southeast Asia. Sub-Saharan Africa has the lowest percentage of its cultivated lands irrigated. Sprinkler and localized irrigation methods are employed throughout the world and account for about 14 percent of the total area of irrigated land.

Statistics on drainage improvements are sparser than statistics on irrigation. It may safely be said that drainage in one form or another is practiced in almost every country of the world. It is now universally accepted that drainage is needed as much on irrigated as on nonirrigated land. Countries such as India that have large-scale river-basin developments planned with irrigation also have companion drainage systems planned so that the land will not be damaged by salt accumulation.

It is almost certain that the land area of the world improved by irrigation and drainage will continue to increase because these practices are two of the most elemental means of reclaiming and improving agricultural lands.

Harvesting

Harvesting is the act of removing a crop from where it was growing and moving it to a more secure location for processing, consumption, or storage. Some root crops and tree fruit can be left in the field or orchard and harvested as needed, but most crops reach a period of maximum quality—that is, they ripen or mature—and will deteriorate if left exposed to the elements. While the major factor determining the time of harvest is the maturity of the crop, other factors such as weather, availability of harvest equipment, pickers, packing and storage facilities, and transport are important considerations.

Economic and marketing issues are often even more important than considerations of maturity in deciding when to harvest a commodity. Before the crop can be harvested, the grower must be sure that there is a demand for the crop and that the price is sufficient to make harvesting the crop profitable. If the price is less than adequate to cover the costs of production, harvesting, and marketing, growers are faced with the difficult decision whether to harvest and store the crop, to wait for a better market, or to cut their losses and leave the crop in the field.

The Process of Harvesting

Harvesting can be separated into three steps. The plant part of interest must be identified, detached from the rest of the plant, and then collected in a container suitable for transport from the field. The harvesting of all the major agronomic crops (grains of cereals and legumes) has been mechanized. The resistance of dried cereal and legume seeds (for example, corn, rice, wheat, and soybeans) to physical damage allows the first and second steps to be combined in a threshing machine or combine that separates the seeds from the rest of the harvested plant. The grain (seeds) is then loaded in bulk containers and transported to silos for additional cleaning, grading, fumigation, and temporary storage.

In contrast to the dry grains and legumes, most horticultural crops (fleshy fruits such as apples and tomatoes, ornamentals, and vegetables) are hand harvested for the fresh market. Some mechanical harvesting is done, but the damage incurred is usually so severe that the fruits and vegetables are only fit for processing. Some fruits (for example, apple, pear, and tomato) evolved to ensure seed distribution by enticing animals to eat the attractive (in appearance, aroma, and flavor) fruit. Therefore, they are almost always attached to the plant by a specialized structure that contains an "abscission zone" which permits the ripe fruit to be easily separated from the tree. Other fruit, such as bananas, citrus, peppers, and all vegetables (for example, cabbage, broccoli, carrots, lettuce, and squash) do not develop an abscission zone and must be cut from the plant.

Only human beings have the unique combination of eyes, brain, and hands that permits the rapid identification and harvest of delicate and perishable materials with minimal loss and bruising. Some mechanical aids such as cherry pickers, ladders, picking bags or baskets, stem clippers, and wheelbarrows are used by harvesters to make their work easier, faster, and safer. Harvesters can be trained to select only those fruits or vegetables that are of the correct maturity, size, or shape, thus greatly reducing the amount of material that must be removed on the sorting and grading line in the packing shed. In fact, many vegetables and berries are harvested directly into retail containers without further sizing or grading. Most other horticultural crops are harvested into field bins that are taken to packing sheds where the commodities are cleaned, sorted, graded, inspected, packed, cooled, and stored before being transported to regional markets.

Agricultural Land

Agricultural land is defined as the land area that is either arable, under permanent crops, or under permanent pastures. Arable land includes land under temporary crops such as cereals, temporary meadows for mowing or for pasture, land under market or kitchen gardens, and land temporarily fallow. Land abandoned as a result of shifting cultivation is excluded. Land under permanent crops is cultivated with crops that occupy the land for long periods and need not be replanted after each harvest, such as orchards and vineyards. This category excludes land under trees grown for wood or timber. Permanent pasture is land used for five or more years for forage, including natural and cultivated crops. This indicator is presented as a total and per type of agricultural land and is measured in hectares and in percentage. Other agro-environmental indicators include Organic farmland and Transgenic cropland.

This sense of "agricultural land" thus includes a great deal of land not devoted to agricultural use. The land actually under annually-replanted crops in any given year is instead said to constitute "sown land" or "cropped land". "Permanent cropland" includes forested plantations used to harvest coffee, rubber, or fruit but not tree farms or proper forests used for wood or timber. Land able to be used for farming is called "cultivable land". Farmland, meanwhile, is used variously in reference to all agricultural land, to all cultivable land, or just to the newly restricted sense of "arable land". Depending upon its use of artificial irrigation, the FAO's "agricultural land" may be divided into irrigated and non-irrigated land.

In the context of zoning, agricultural land or agriculturally-zoned land refers to plots that are *permitted* to be used for agricultural activities, without regard to its present use or even suitability. In some areas, agricultural land is protected so that it can be farmed without any threat of development. The Agricultural Land Reserve in British Columbia in Canada, for instance, requires approval from its Agricultural Land Commission before its lands can be removed or subdivided.

All ecosystems modified or created by man specifically to grow or raise biological products for human consumption or use. This includes cropland, pasture, orchards, groves, vineyards, nurseries, ornamental horticultural areas, and confined feeding areas.

Agricultural Land Categories

- Confined Feeding Operations: Those ecosystems that are modified or created by man to serve as large, specialized livestock production enterprises in which large animal populations are restricted to relatively small areas. This include beef cattle feedlots, dairy operations with confined feeding, large poultry farms, and hog feedlots. This does not include shipping corrals and other temporary holding facilities.
- Cropland and Pasture: Those ecosystems that are modified or created by man to produce agricultural crops (e.g., corn, wheat, soybean) and/or to be used for pasture. This includes harvested cropland, cultivated summer-fallow and idle cropland, land on which crop failure occurs, cropland in soil-improvement grasses and legumes, cropland used only for pasture in rotation with crops, and pasture on land more or less permanently used for that purpose.
- Orchards and Vinyards: Those ecosystems that are modified or created by man to produce the various fruit and nut crops. This includes orchards, groves, vineyards, nurseries, and ornamental horticulture areas. This includes floricultural and seed-and-sod areas and some greenhouses, as well as tree nurseries that provide seedlings for plantation forestry.
- Other Agricultural Land: Those ecosystems that are modified or created by man to produce food and fiber, but do not fall into the other categories of agricultural land (cropland and pasture; orchards, groves, vineyards, nurseries and ornamental horticultural areas; confined feeding operations). This includes farmsteads, holdings for livestock such as corrals, breeding and training facilities on horse farms, farm lanes and roads, ditches and canals, small farm ponds, and similar uses.

Fertilizer

Fertilizer is a natural or artificial substance containing the chemical elements that improve growth and productiveness of plants. Fertilizers enhance the natural fertility of the soil or replace the chemical elements taken from the soil by previous crops.

The use of manure and composts as fertilizers is probably almost as old as agriculture. Modern chemical fertilizers include one or more of the three elements that are most important in plant nutrition: nitrogen, phosphorus, and potassium. Of secondary importance are the elements sulfur, magnesium, and calcium.

Most nitrogen fertilizers are obtained from synthetic ammonia; this chemical compound (NH_3) is used either as a gas or in a water solution, or it is converted into salts such as ammonium sulfate, ammonium nitrate, and ammonium phosphate, but packinghouse wastes, treated garbage, sewage, and manure are also common sources of it. Phosphorus fertilizers include calcium phosphate derived from phosphate rock or bones. The more soluble superphosphate and triple superphosphate preparations are obtained by the treatment of calcium phosphate with sulfuric and phosphoric acid, respectively. Potassium fertilizers, namely potassium chloride and potassium sulfate, are mined from potash deposits. Mixed fertilizers contain more than one of the three major nutrients—nitrogen, phosphorus, and potassium. Mixed fertilizers can be formulated in hundreds of ways.

On modern farms a variety of machines are used to apply synthetic fertilizer in solid, gaseous, or liquid form. One type distributes anhydrous ammonia, a liquid under pressure, which becomes a nitrogenous gas when freed from pressure as it enters the soil. A metering device operates valves to release the liquid from the tank. Solid-fertilizer distributors have a wide hopper, with holes in the bottom; distribution is effected by various means, such as rollers, agitators, or endless chains traversing the hopper bottom. Broadcast distributors have a tub-shaped hopper from which the material falls onto revolving disks that distribute it in a broad swath.

Fertilizers enhance the growth of plants. This goal is met in two ways, the traditional one being additives that provide nutrients. The second mode by which some fertilizers act is to enhance the effectiveness of the soil by modifying its water retention and aeration.

- Three main macronutrients:
  - Nitrogen (N): leaf growth;
  - Phosphorus (P): Development of roots, flowers, seeds, fruit;
  - Potassium (K): Strong stem growth, movement of water in plants, promotion of flowering and fruiting.
- Three secondary macronutrients: calcium (Ca), magnesium (Mg), and sulfur (S).
- Micronutrients: copper (Cu), iron (Fe), manganese (Mn), molybdenum (Mo), zinc (Zn), boron (B). Of occasional significance are silicon (Si), cobalt (Co), and vanadium (V).

The nutrients required for healthy plant life are classified according to the elements, but the elements are not used as fertilizers. Instead compounds containing these elements are the basis of fertilizers. The macro-nutrients are consumed in larger quantities and are present in plant tissue in quantities from 0.15% to 6.0% on a dry matter (DM) (0% moisture) basis. Plants are made up of four main elements: hydrogen, oxygen, carbon, and nitrogen. Carbon, hydrogen and oxygen are widely available as water and carbon dioxide. Although nitrogen makes up most of the atmosphere, it is in a form that is unavailable to plants. Nitrogen is the most important fertilizer since nitrogen is present in proteins, DNA and other components (e.g., chlorophyll). To be nutritious to plants, nitrogen must be made available in a "fixed" form. Only some bacteria and their host plants (notably legumes) can fix atmospheric nitrogen (N_2) by converting it to ammonia. Phosphate is required for the production of DNA and ATP, the main energy carrier in cells, as well as certain lipids.

Micronutrients are consumed in smaller quantities and are present in plant tissue on the order of parts-per-million (ppm), ranging from 0.15 to 400 ppm DM, or less than 0.04% DM. These elements are often present at the active sites of enzymes that carry out the plant's metabolism. Because these elements enable catalysts (enzymes) their impact far exceeds their weight percentage.

Classification

Fertilizers are classified in several ways. They are classified according to whether they provide a single nutrient (e.g., K, P, or N), in which case they are classified as "straight fertilizers." "Multinutrient fertilizers" (or "complex fertilizers") provide two or more nutrients, for example N and P. Fertilizers are also sometimes classified as inorganic versus organic. Inorganic fertilizers exclude carbon-containing materials except ureas. Organic fertilizers are usually (recycled) plant- or animal-derived matter. Inorganic are sometimes called synthetic fertilizers since various chemical treatments are required for their manufacture.

Single Nutrient "Straight" Fertilizers

The main nitrogen-based straight fertilizer is ammonia or its solutions. Ammonium nitrate (NH_4NO_3) is also widely used. Urea is another popular source of nitrogen, having the advantage that it is solid and non-explosive, unlike ammonia and ammonium nitrate, respectively. A few percent of the nitrogen fertilizer market (4% in 2007) has been met by calcium ammonium nitrate ($Ca(NO_3)_2 \cdot NH_4 \cdot 10H_2O$).

The main straight phosphate fertilizers are the superphosphates. "Single superphosphate" (SSP) consists of 14–18% P_2O_5, again in the form of $Ca(H_2PO_4)_2$, but also phosphogypsum ($CaSO_4 \cdot 2H_2O$). Triple superphosphate (TSP) typically consists of 44-48% of P_2O_5 and no gypsum. A mixture of single superphosphate and triple superphosphate is called double superphosphate. More than 90% of a typical superphosphate fertilizer is water-soluble.

The main potassium-based straight fertilizer is Muriate of Potash (MOP). Muriate of Potash consists of 95-99% KCl, and is typically available as 0-0-60 or 0-0-62 fertilizer.

Multinutrient Fertilizers

These fertilizers are common. They consist of two or more nutrient components.

Binary (NP, NK, PK) Fertilizers

Major two-component fertilizers provide both nitrogen and phosphorus to the plants. These are called NP fertilizers. The main NP fertilizers are monoammonium phosphate (MAP) and diammonium phosphate (DAP). The active ingredient in MAP is $NH_4H_2PO_4$. The active ingredient in DAP is $(NH_4)_2HPO_4$. About 85% of MAP and DAP fertilizers are soluble in water.

NPK Fertilizers

NPK fertilizers are three-component fertilizers providing nitrogen, phosphorus, and potassium.

NPK rating is a rating system describing the amount of nitrogen, phosphorus, and potassium in a fertilizer. NPK ratings consist of three numbers separated by dashes (e.g., 10-10-10 or 16-4-8) describing the chemical content of fertilizers. The first number represents the percentage of nitrogen in the product; the second number, P_2O_5; the third, K_2O. Fertilizers do not actually contain P_2O_5 or K_2O, but the system is a conventional shorthand for the amount of the phosphorus (P) or potassium (K) in a fertilizer. A 50-pound (23 kg) bag of fertilizer labeled 16-4-8 contains 8 lb (3.6 kg) of nitrogen (16% of the 50 pounds), an amount of phosphorus equivalent to that in 2 pounds of P_2O_5 (4% of 50 pounds), and 4 pounds of K_2O (8% of 50 pounds). Most fertilizers are labeled according to this N-P-K convention, although Australian convention, following an N-P-K-S system, adds a fourth number for sulfur, and uses elemental values for all values including P and K.

Micronutrients

The main micronutrients are molybdenum, zinc, boron, and copper. These elements are provided as water-soluble salts. Iron presents special problems because it converts to insoluble (bio-unavailable) compounds at moderate soil pH and phosphate concentrations. For this reason, iron is often administered as a chelate complex, e.g., the EDTA derivative. The micronutrient needs depend on the plant and the environment. For example, sugar beets appear to require boron, and legumes require cobalt, while environmental conditions such as heat or drought make boron less available for plants.

Nitrogen Fertilizers

Nitrogen fertilizers are made from ammonia (NH_3), which is sometimes injected into the ground directly. The ammonia is produced by the Haber-Bosch process. In this energy-intensive process, natural gas (CH_4) usually supplies the hydrogen, and the nitrogen (N_2) is derived from the air. This ammonia is used as a feedstock for all other nitrogen fertilizers, such as anhydrous ammonium nitrate (NH_4NO_3) and urea ($CO(NH_2)_2$).

Deposits of sodium nitrate ($NaNO_3$) (Chilean saltpeter) are also found in the Atacama desert in Chile and was one of the original (1830) nitrogen-rich fertilizers used. It is still mined for fertilizer. Nitrates are also produced from ammonia by the Ostwald process.

Phosphate Fertilizers

All phosphate fertilizers are obtained by extraction from minerals containing the anion PO_4^{3-}. In rare cases, fields are treated with the crushed mineral, but most often more soluble salts are produced by chemical treatment of phosphate minerals. The most popular phosphate-containing minerals are referred to collectively as phosphate rock. The main minerals are fluorapatite $Ca_5(PO_4)_3F$ (CFA) and hydroxyapatite $Ca_5(PO_4)_3OH$. These minerals are converted to water-soluble phosphate salts by treatment with sulfuric (H_2SO_4) or phosphoric acids (H_3PO_4). The large production of sulfuric acid as an industrial chemical is primarily due to its use as cheap acid in processing phosphate rock into phosphate fertilizer. The global primary uses for both sulfur and phosphorus compounds relate to this basic process.

In the nitrophosphate process or Odda process (invented in 1927), phosphate rock with up to a 20% phosphorus (P) content is dissolved with nitric acid (HNO_3) to produce a mixture of phosphoric acid (H_3PO_4) and calcium nitrate ($Ca(NO_3)_2$). This mixture can be combined with a potassium fertilizer to produce a *compound fertilizer* with the three macronutrients N, P and K in easily dissolved form.

Potassium Fertilizers

Potash is a mixture of potassium minerals used to make potassium (chemical symbol: K) fertilizers. Potash is soluble in water, so the main effort in producing this nutrient from the ore involves some purification steps; e.g., to remove sodium chloride (NaCl) (common salt). Sometimes potash is referred to as K_2O, as a matter of convenience to those describing the potassium content. In fact, potash fertilizers are usually potassium chloride, potassium sulfate, potassium carbonate, or potassium nitrate.

Compound Fertilizers

Compound fertilizers, which contain N, P, and K, can often be produced by mixing straight fertilizers. In some cases, chemical reactions occur between the two or more components. For example, monoammonium and diammonium phosphates, which provide plants with both N and P, are produced by neutralizing phosphoric acid (from phosphate rock) and ammonia:

$$NH_3 + H_3PO_4 \rightarrow (NH_4)H_2PO_4$$

$$2\ NH_3 + H_3PO_4 \rightarrow (NH_4)_2HPO_4$$

Organic Fertilizers

"Organic fertilizers" can describe those fertilizers with an organic — biologic — origin—that is, fertilizers derived from living or formerly living materials. Organic fertilizers can also describe commercially available and frequently packaged products that strive to follow the expectations and restrictions adopted by "organic agriculture" and "environmentally friendly" gardening — related systems of food and plant production that significantly limit or strictly avoid the use of synthetic fertilizers and pesticides. The "organic fertilizer" products typically contain both some organic materials as well as acceptable additives such as nutritive rock powders, ground sea shells (crab, oyster, etc.), other prepared products such as seed meal or kelp, and cultivated microorganisms and derivatives.

Compost bin for small-scale production of organic fertilizer.

Fertilizers of an organic origin (the first definition) include animal wastes, plant wastes from agriculture, compost, and treated sewage sludge (biosolids). Beyond manures, animal sources can include products from the slaughter of animals — bloodmeal, bone meal, feather meal, hides, hoofs, and horns all are typical components. Organically derived materials available to industry such as sewage sludge may not be acceptable components of organic farming and gardening, because of factors ranging from residual contaminants to public perception. On the other hand, marketed "organic fertilizers" may include, and promote, processed organics *because* the materials have consumer appeal. No matter the definition nor composition, most of these products contain less concentrated nutrients, and the nutrients are not as easily quantified. They can offer soil-building advantages as well as be appealing to those who are trying to farm/garden more "naturally".

A large commercial compost operation.

In terms of volume, peat is the most widely used packaged organic soil amendment. It is an immature form of coal and improves the soil by aeration and absorbing water but confers no nutritional value to the plants. Coir, (derived from coconut husks), bark, and sawdust when added to soil all act similarly (but not identically) to peat and are also considered organic soil amendments - or texturizers - because of their limited nutritive inputs. Some organic additives can have a reverse effect on nutrients — fresh sawdust can consume soil nutrients as it breaks down, and may lower soil pH — but these same organic texturizers (as well as compost, etc.) may increase the availability of nutrients through improved cation exchange, or through increased growth of microorganisms that in turn increase availability of certain plant nutrients. Organic fertilizers such as composts and manures may be distributed locally without going into industry production, making actual consumption more difficult to quantify.

Application

Fertilizers are commonly used for growing all crops, with application rates depending on the soil fertility, usually as measured by a soil test and according to the particular crop. Legumes, for example, fix nitrogen from the atmosphere and generally do not require nitrogen fertilizer.

Liquid vs. Solid

Fertilizers are applied to crops both as solids and as liquid. About 90% of fertilizers are applied as solids. The most widely used solid inorganic fertilizers are urea, diammonium phosphate and potassium chloride. Solid fertilizer is typically granulated or powdered. Often solids are available as prills, a solid globule. Liquid fertilizers comprise anhydrous ammonia, aqueous solutions of ammonia, aqueous solutions of ammonium nitrate or urea. These concentrated products may be diluted with water to form a concentrated liquid fertilizer (e.g., UAN). Advantages of liquid fertilizer are its more rapid effect and easier coverage. The addition of fertilizer to irrigation water is called "fertigation".

Slow- and Controlled-release Fertilizers

Slow- and controlled-release involve only 0.15% (562,000 tons) of the fertilizer market. Their utility stems from the fact that fertilizers are subject to antagonistic processes. In addition to their providing the nutrition to plants, excess fertilizers can be poisonous to the same plant. Competitive with the uptake by plants is the degradation or loss of the fertilizer. Microbes degrade many fertilizers, e.g., by immobilization or oxidation. Furthermore, fertilizers are lost by evaporation or leaching. Most slow-release fertilizers are derivatives of urea, a straight fertilizer providing nitrogen. Isobutylidenediurea ("IBDU") and urea-formaldehyde slowly convert in the soil to free urea, which is rapidly uptaken by plants. IBDU is a single compound with the formula $(CH_3)_2CHCH(NHC(O)NH_2)_2$ whereas the urea-formaldehydes consist of mixtures of the approximate formula $(HOCH_2NHC(O)NH)_nCH_2$.

Besides being more efficient in the utilization of the applied nutrients, slow-release technologies also reduce the impact on the environment and the contamination of the subsurface water. Slow-release fertilizers (various forms including fertilizer spikes, tabs, etc.) which reduce the problem of "burning" the plants due to excess nitrogen. Polymer coating of fertilizer ingredients gives tablets and spikes a 'true time-release' or 'staged nutrient release' (SNR) of fertilizer nutrients.

Controlled release fertilizers are traditional fertilizers encapsulated in a shell that degrades at a specified rate. Sulfur is a typical encapsulation material. Other coated products use thermoplastics (and sometimes ethylene-vinyl acetate and surfactants, etc.) to produce diffusion-controlled release of urea or other fertilizers. "Reactive Layer Coating" can produce thinner, hence cheaper, membrane coatings by applying reactive monomers simultaneously to the soluble particles. "Multicote" is a process applying layers of low-cost fatty acid salts with a paraffin topcoat.

Foliar Application

Foliar fertilizers are applied directly to leaves. The method is almost invariably used to apply water-soluble straight nitrogen fertilizers and used especially for high value crops such as fruits.

Fertilizer burn.

Chemicals that Affect Nitrogen Uptake

Various chemicals are used to enhance the efficiency of nitrogen-based fertilizers. In this way farmers can limit the polluting effects of nitrogen run-off. Nitrification inhibitors (also known as nitrogen stabilizers) suppress the conversion of ammonia into nitrate, an anion that is more prone to leaching. 1-Carbamoyl-3-methylpyrazole (CMP), dicyandiamide, nitrapyrin (2-chloro-6-trichloromethylpyridine) and 3,4-Dimethylpyrazole phosphate (DMPP) are popular. Urease inhibitors are used to slow the hydrolytic conversion of urea into ammonia, which is prone to evaporation as well as nitrification. The conversion of urea to ammonia catalyzed by enzymes called ureases. A popular inhibitor of ureases is N-(n-butyl)thiophosphoric triamide (NBPT).

Overfertilization

Careful fertilization technologies are important because excess nutrients can be detrimental. Fertilizer burn can occur when too much fertilizer is applied, resulting in damage or even death of the plant. Fertilizers vary in their tendency to burn roughly in accordance with their salt index.

Environmental Effects

Runoff of soil and fertilizer during a rain storm.

Use of fertilizers are beneficial in providing nutrients to plants although they have some negative environmental effects. The large growing consumption of fertilizers can affect soil, surface water, and groundwater due to dispersion of mineral use.

Water

Phosphorus and nitrogen fertilizers when commonly used have major environmental effects. This is due to high rainfalls causing the fertilizers to be washed into waterways. Agricultural run-off is a major contributor to the eutrophication of fresh water bodies. For example, in the US, about half of all the lakes are eutrophic. The main contributor to eutrophication is phosphate, which is normally a limiting nutrient; high concentrations promote the growth of cyanobacteria and algae, the demise of which consumes oxygen. Cyanobacteria blooms ('algal blooms') can also produce harmful toxins that can accumulate in the food chain, and can be harmful to humans.

The nitrogen-rich compounds found in fertilizer runoff are the primary cause of serious oxygen depletion in many parts of oceans, especially in coastal zones, lakes and rivers. The resulting lack of dissolved oxygen greatly reduces the ability of these areas to sustain oceanic fauna. The number of oceanic dead zones near inhabited coastlines are increasing. As of 2006, the application of nitrogen fertilizer is being increasingly controlled in northwestern Europe and the United States. If eutrophication *can* be reversed, it may take decades before the accumulated nitrates in groundwater can be broken down by natural processes.

Nitrate Pollution

Only a fraction of the nitrogen-based fertilizers is converted to plant matter. The remainder accumulates in the soil or is lost as run-off. High application rates of nitrogen-containing fertilizers combined with the high water solubility of nitrate leads to increased runoff into surface water as well as leaching into groundwater, thereby causing groundwater pollution. The excessive use of nitrogen-containing fertilizers (be they synthetic or natural) is particularly damaging, as much of the nitrogen that is not taken up by plants is transformed into nitrate which is easily leached.

Nitrate levels above 10 mg/L (10 ppm) in groundwater can cause 'blue baby syndrome' (acquired methemoglobinemia). The nutrients, especially nitrates, in fertilizers can cause problems for natural habitats and for human health if they are washed off soil into watercourses or leached through soil into groundwater.

Soil

Acidification

Nitrogen-containing fertilizers can cause soil acidification when added. This may lead to decrease in nutrient availability which may be offset by liming.

Accumulation of Toxic Elements

Cadmium

The concentration of cadmium in phosphorus-containing fertilizers varies considerably and can be problematic. For example, mono-ammonium phosphate fertilizer may have a cadmium content of as low as 0.14 mg/kg or as high as 50.9 mg/kg. The phosphate rock used in their manufacture can contain as much as 188 mg/kg cadmium (examples are deposits on Nauru and the Christmas islands). Continuous use of high-cadmium fertilizer can contaminate soil and plants. Limits to the cadmium content of phosphate fertilizers has been considered by the European Commission. Producers of phosphorus-containing fertilizers now select phosphate rock based on the cadmium content.

Fluoride

Phosphate rocks contain high levels of fluoride. Consequently, the widespread use of phosphate fertilizers has increased soil fluoride concentrations. It has been found that food contamination from fertilizer is of little concern as plants accumulate little fluoride from the soil; of greater concern is the possibility of fluoride toxicity to livestock that ingest contaminated soils. Also of possible concern are the effects of fluoride on soil microorganisms.

Radioactive Elements

The radioactive content of the fertilizers varies considerably and depends both on their concentrations in the parent mineral and on the fertilizer production process. Uranium-238 concentrations can range from 7 to 100 pCi/g in phosphate rock and from 1 to 67 pCi/g in phosphate fertilizers. Where high annual rates of phosphorus fertilizer are used, this can result in uranium-238 concentrations in soils and drainage waters that are several times greater than are normally present. However, the impact of these increases on the risk to human health from radinuclide contamination of foods is very small (less than 0.05 mSv/y).

Other Metals

Steel industry wastes, recycled into fertilizers for their high levels of zinc (essential to plant growth), wastes can include the following toxic metals: lead arsenic, cadmium, chromium, and nickel. The most common toxic elements in this type of fertilizer are mercury, lead, and arsenic. These potentially harmful impurities can be removed; however, this significantly increases cost. Highly pure fertilizers are widely available and perhaps best known as the highly water-soluble fertilizers containing blue dyes used around households, such as Miracle-Gro. These highly water-soluble fertilizers are used in the plant nursery business and are available in larger packages at significantly less cost than retail quantities. Some inexpensive retail granular garden fertilizers are made with high purity ingredients.

Trace Mineral Depletion

Attention has been addressed to the decreasing concentrations of elements such as iron, zinc, copper and magnesium in many foods over the last 50–60 years. Intensive farming practices, including the use of synthetic fertilizers are frequently suggested as reasons for these declines and organic farming is often suggested as a solution. Although improved crop yields resulting from NPK fertilizers are known to dilute the concentrations of other nutrients in plants, much of the measured decline can be attributed to the use of progressively higher-yielding crop varieties which produce foods with lower mineral concentrations than their less productive ancestors. It is, therefore, unlikely that organic farming or reduced use of fertilizers will solve the problem; foods with high nutrient density are posited to be achieved using older, lower-yielding varieties or the development of new high-yield, nutrient-dense varieties.

Fertilizers are, in fact, more likely to solve trace mineral deficiency problems than cause them: In Western Australia deficiencies of zinc, copper, manganese, iron and molybdenum were identified as limiting the growth of broad-acre crops and pastures in the 1940s and 1950s. Soils in Western Australia are very old, highly weathered and deficient in many of the major nutrients and trace elements. Since this time these trace elements are routinely added to fertilizers used in agriculture in this state. Many other soils around the world are deficient in zinc, leading to deficiency in both plants and humans, and zinc fertilizers are widely used to solve this problem.

Changes in Soil Biology

High levels of fertilizer may cause the breakdown of the symbiotic relationships between plant roots and mycorrhizal fungi.

Energy Consumption and Sustainability

In the US in 2004, 317 billion cubic feet of natural gas were consumed in the industrial production of ammonia, less than 1.5% of total U.S. annual consumption of natural gas. A 2002 report suggested that the production of ammonia consumes about 5% of global natural gas consumption, which is somewhat under 2% of world energy production.

Ammonia is produced from natural gas and air. The cost of natural gas makes up about 90% of the cost of producing ammonia. The increase in price of natural gases over the past decade, along with other factors such as increasing demand, have contributed to an increase in fertilizer price.

Contribution to Climate Change

The greenhouse gases carbon dioxide, methane and nitrous oxide are produced during the manufacture of nitrogen fertilizer. The effects can be combined into an equivalent amount of carbon dioxide. The amount varies according to the efficiency of the process. The figure for the United Kingdom is over 2 kilogrammes of carbon dioxide equivalent for each kilogramme of ammonium nitrate. Nitrogen fertilizer can be converted by soil bacteria to nitrous oxide, a greenhouse gas.

Atmosphere

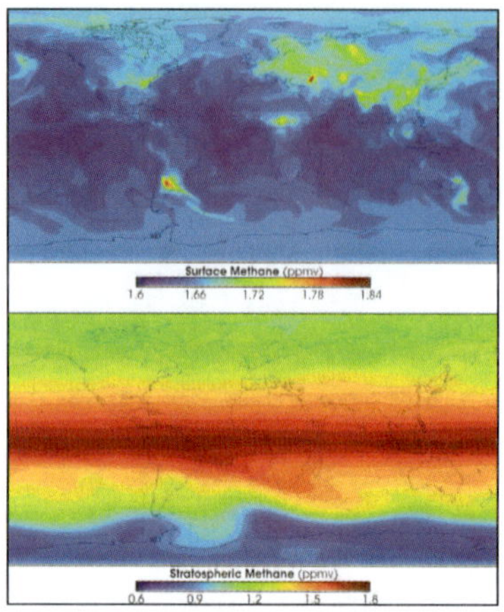

Global methane concentrations (surface and atmospheric).

Through the increasing use of nitrogen fertilizer, which was used at a rate of about 110 million tons (of N) per year in 2012, adding to the already existing amount of reactive nitrogen, nitrous oxide (N_2O) has become the third most important greenhouse gas after carbon dioxide and methane. It has a global warming potential 296 times larger than an equal mass of carbon dioxide and it also contributes to stratospheric ozone depletion. By changing processes and procedures, it is possible to mitigate some, but not all, of these effects on anthropogenic climate change.

Methane emissions from crop fields (notably rice paddy fields) are increased by the application of ammonium-based fertilizers. These emissions contribute to global climate change as methane is a potent greenhouse gas.

Regulation

In Europe problems with high nitrate concentrations in run-off are being addressed by the European Union's Nitrates Directive. Within Britain, farmers are encouraged to manage their land more sustainably in 'catchment-sensitive farming'. In the US, high concentrations of nitrate and phosphorus in runoff and drainage water are classified as non-point source pollutants due to their diffuse origin; this pollution is regulated at state level. Oregon and Washington, both in the United States, have fertilizer registration programs with on-line databases listing chemical analyses of fertilizers.

In China, there have been regulations implemented by the government that want to control N fertilizers being used in farming. In 2008, Chinese governments have started to partially withdraw fertilizer subsidies, which also include contributions to fertilizer transportation, electricity and natural gas use in the industry. Because of this, professional farmers who run large-scale farms have already used less fertilizers since then under the fertilizer prices went up. If large-scale farms keep reducing their use of fertilizer subsidies, they have no choice but to optimize the fertilizer they have which would therefore gain an increase in both grain yield and profit.

Two types of agricultural management practices include organic agriculture and conventional agriculture. The former encourages soil fertility using local resources to maximize efficiency. Organic agriculture avoids synthetic agrochemicals. Conventional agriculture uses all the components that organic agriculture does not use.

Agricultural Machinery

Agricultural Machinery devices used to till soil and to plant, cultivate, and harvest crops. Since ancient times, when cultures first began cultivating plants, people have used tools to help them grow and harvest crops. They used pointed tools to dig and keep soil loosened, and sharp, knifelike objects to harvest ripened crops. Modifications of these early implements led to the development of small hand tools that are still used in small-scale gardening, such as the spade, hoe, rake, trowel, and scythe, and larger implements, such as plows and larger rakes that are drawn by humans, animals, or simple machines.

Modern large agricultural implements, adapted to large-scale farming methods, are usually powered by diesel- or gasoline-fueled internal-combustion engines. The most important implement of modern agriculture is the tractor. It provides locomotion for many other implements and can furnish power, via its power shaft, for the operation of machines drawn behind the tractor. The power shafts of tractors can also be set up to drive belts that operate equipment such as feed grinders, pumps, and electric-power generators. Small implements, such as portable irrigators, are often powered by individual motors.

Implements for Growing Crops

Many types of implements have been developed for the activities involved in growing crops. These activities include breaking ground, planting, weeding, fertilizing, and combatting pests.

Ground is broken by plows to prepare the seedbed. A plow consists of a bladelike plowshare that cuts under, then lifts, turns, and pulverizes the soil. Modern tractor plows are usually equipped with two or more plowshares so that a wide area of ground can be broken at a single sweep. Harrows are used to smooth the plowed land and sometimes to cover seeds and fertilizer with earth. The disk harrow, which has curved, sharp-edged steel disks, is used mainly to cut up crop residues before plowing and to bury weeds during seedbed preparation. Rollers with V-shaped wheels break up clods of soil to improve the aeration of the soil and its capacity for taking in water.

Some cereal crops are still planted by broadcasting seeds–that is, by scattering the seeds over a wide area. Machines for broadcasting usually consist of a long seedbox mounted on wheels and equipped with an agitator to distribute the seeds. Broadcast seeds are not always covered by a uniform or sufficient depth of soil, so seeding is more often done with drills, which produce continuous furrows of uniform depth. Specialized implements called planters are necessary for sowing crops that are planted in rows, such as corn. Corn planters and other similar machines have a special feed wheel that picks up small quantities of grain or separate kernels and places them in the ground.

Fertilizer can be distributed during the winter or shortly before seeding time. Commercial fertilizers are commonly distributed, along with seeds, by drills and planters. Manure is distributed most efficiently by a manure spreader, which is a wagon equipped with a bottom conveyor to carry the fertilizer back to a beater attachment, which disintegrates it and then scatters it on the ground.

After crops have begun to grow, a cultivator is used to destroy weeds and loosen and aerate the soil. A flame weeder, which produces a hot-air blast, can be used to destroy weeds growing around crops, such as cotton, that have stems of tough bark. The weeds are vulnerable to the hot air, but the tough stems protect the crops from damage. Chemical herbicides applied in the form of a spray or as granules are used extensively for destroying weeds.

Insecticides are applied to soil and crops in the form of granules, dust, or liquid sprays. A variety of mechanical spraying and dusting equipment is used to spread chemicals on crops and fields; the machinery may be self-powered, or drawn and powered by a tractor. In areas where large crops of vegetables and grain are grown, airplanes are sometimes used to dust or spray pesticides.

Chemical pesticides are used in nearly all modern farming operations. However, increasing concern over the harmful effects that pesticides may have on the environment has led to the use of alternative forms of pest control. For example, farmers use crop rotation to prevent pests that feed on a certain crop from becoming entrenched and infesting the field. Also, certain pests are controlled by introducing an organism that damages or kills the pests, but leaves the crops unharmed. Finally, scientists genetically engineer crops to be more resistant to troublesome pests.

Implements for Harvesting Crops

Most cereal crops are harvested by using a combine–a machine that removes the fruiting heads, beats off the grain kernels, and cleans the grain as the combine moves through the fields. The cleaned grain is accumulated in an attached grain tank.

Corn is harvested by a combine or a machine called a corn picker. As the corn picker moves along the rows, the ears are picked from the stalks and are husked. The ears are then transferred either to a sheller, which removes the kernels from the ear, or to a wagon trailing behind the machine.

Hay harvesting usually requires several steps. First, the hay is cut close to the ground with a mower. After drying in the sun, most hay is baled. In baling, the pickup baler lifts the hay to a conveyor that carries it to a baling chamber, which compresses the hay into bales weighing up to 57 kg (125 lb) and ties each bale with heavy twine or wire. A machine called a field chopper cuts down green hay or field-cured hay for use as animal feed. After being cut down, the hay is stored in a silo and allowed to ferment; this type of animal feed is nutritious and resistant to spoilage. Alfalfa and other legume hay is harvested in some areas with a hay cuber. This machine cuts the plants close to the ground and, after field curing, chops them into a fine mash and compresses the mash into cubes that are more easily shipped and stored than are bales.

Specialized machinery is used to harvest large root crops such as potatoes and sugar beets. Mechanical cotton pickers and strippers are used in harvesting nearly all of the cotton grown in the United

States. Mechanical pickers have rotating spindles that twist the cotton fiber from the boll. Before picking, the leaves of the cotton plant are removed by means of a chemical defoliant spray. Light-boll, stormproof cotton is harvested by strippers that comb or brush the cotton from the plant and lift it into a trailed wagon. More efficient mechanical pickers continue to be developed.

Specialized machines are also used to harvest fruits and vegetables. Some mechanical fruit pickers that are used to harvest deciduous tree fruits, such as plums, cherries, and apricots shake the fruit tree, causing the fruit to fall onto a raised catching frame that surrounds the tree. Nut crops can also be harvested in this manner. In addition, plant breeders use modern methods such as genetic engineering to develop varieties of fruits and vegetables that are tougher and hardier for easy harvesting by machines. For example, a variety of tomato has been bred for a tougher skin that reduces bruising.

In addition to the kinds of agricultural machinery being used on large modern farms, a wide range of automated devices became available to farmers through the revolution in electronics. Today, an increasing number of farmers use personal computers to keep records, manage their farms' business, and connect to information centers that can help them solve the problems they confront in the operation of their farms.

Practical Significance

Use of agricultural machinery substantially reduces the amount of human labor needed for raising crops. The average amount of labor required per hectare to produce and harvest corn, hay, and cereal crops has fallen to less than a fourth of what was required only a few decades ago.

References

- Crop-selection: cropsreview.com, Retrieved 21 March , 2019
- Soil-preparation, crop-production-and-management, biology, guides: toppr.com, Retrieved 12 April, 2019
- Agriculture-seeds-selection-sowing, biology: byjus.com, Retrieved 23 June, 2019
- Irrigation, technology: britannica.com, Retrieved 15 May, 2019
- Harvesting, food-and-cooking, food-and-drink, sports-and-everyday-life: encyclopedia.com, Retrieved 20 July, 2019
- Lark, Tyler J.; Meghan Salmon, J.; Gibbs, Holly K. (2015). "Cropland expansion outpaces agricultural and biofuel policies in the United States". Environmental Research Letters. 10 (4): 044003. Bibcode:2015ERL....10d4003L. doi:10.1088/1748-9326/10/4/044003
- Agricultural-land, garland: oecd.org, Retrieved 25 August, 2019
- Agricultural-Land, iwgsdi: nasa.gov, Retrieved 31 July, 2019
- Fertilizer, topic: britannica.com, Retrieved 1 May, 2019
- Trevor Illtyd Williams; Thomas Kingston Derry (1982). A short history of twentieth-century technology c. 1900-c. 1950. Oxford University Press. pp. 134–135. ISBN 978-0-19-858159-8
- Agriculturalmachinery, machinesandtools: colorado.edu, Retrieved 4 Feb, 2019

Chapter 4

Agricultural Soil Science

The study of edaphic conditions in connection with the production of food and fiber is known as agricultural soil science. It is primarily concerned with the study of soil nutrients, soil texture and soil fertility. The chapter closely examines these key concepts of agricultural soil science to provide an extensive understanding of the subject.

Soil is the loose surface material consisting of inorganic particles and organic matter that covers most of the land surface. Soil provides the structural support and the source of water and nutrients for plants used in agriculture. Human use and management of soil and water resources have shaped the development, persistence, decline, and regeneration of human civilizations that are sustained by agriculture. Soil and water are essential natural resources for our domesticated animal- and plant-based food production systems. Although of fundamental importance today, agriculture is a relatively recent human innovation that spread rapidly across the globe only 10,000 to 12,000 years ago, during the Agricultural Revolution. This short, yet highly significant period of time, represents less than 0.3% of the more than four million years of human evolution as bipedal hominids and ultimately Homo sapiens. In agriculturally-based societies during the last ten millennia, humans have developed complex, urban civilizations that have cycled through periods of increasing complexity, awe-inspiring intellectual achievement, persistence for millennia, and, in some instances, perplexing decline. In many cases, stressed, declining civilizations adapted, or reemerged, into new or similar complex cultures. Through such fluctuations, we have remained dependent on a relatively small number of crop and animal species for food, and on integrated soil-water systems that are essential for their production. There is no doubt that our modern human society has developed to the point that we cannot exist without agriculture.

It is clear that agriculture sustains and defines our modern lives, but it is often disruptive of natural ecosystems. This is especially true for plant communities, animal populations, soil systems, and water resources. Understanding, evaluating, and balancing detrimental and beneficial agricultural disturbances of soil and water resources are essential tasks in human efforts to sustain and improve human well-being. Such knowledge influences our emerging ethics of sustainability and responsibility to human populations and ecosystems of the future.

Although agriculture is essential for human food and the stability of complex societies, almost all of our evolution has taken place in small, mobile, kin-based social groups, such as bands and tribes. Before we became sedentary people dependent on agriculture, we were largely dependent on wild plant and animal foods, without managing soil and water resources for food production. Our social evolution has accelerated since the Agricultural Revolution and taken place synergistically with human biological evolution, as we have become dependent on domesticated plants and animals grown purposefully in highly managed, soil-water systems.

Soil Fertility and Crop Growth

The early use of fire to flush out wild game and to clear forested land provided the first major anthropogenic influence on the environment. By burning native vegetation, early humans were able to gain access to herbivores grazing on the savanna and in nearby woodlands, and to suppress the growth of less desirable plant species for those easier to forage and eat. These and other factors (e.g., population pressures, climate change, encouraging/protecting desirable plants), help to lay the groundwork for the Agricultural Revolution and caused a dramatic shift in the interactions between humans and the earth. The shift from hunter-gatherer societies to an agrarian way of life drastically changed the course of human history and irreversibly altered natural nutrient cycling within soils. When humans sowed the first crop seeds at the dawn of the Neolithic Period, the soil provided plant-essential nutrients and served as the foundation for human agriculture.

Plant Nutrients

Throughout Earth's history, natural cycling of nutrients has occurred from the soil to plants and animals, and then back to the soil, primarily through decomposition of biomass. This cycling helps to maintain the essential nutrients required for plant growth in the soil. Complex nutrient cycles incorporate a range of physical, chemical, and — most importantly — biological processes to trace the fate of specific plant nutrients (e.g., N, P, C, S) in the environment.

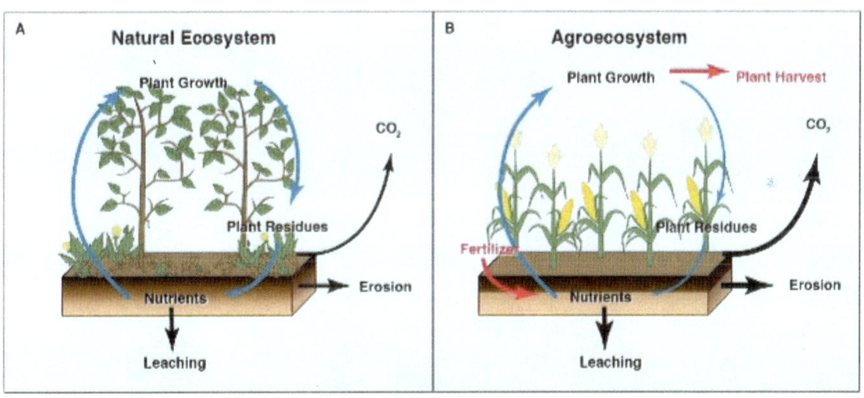

Simplified nutrient cycling: (a) a natural ecosystem and (b) an agroecosystem.

The thickness of arrows correspond to the relative amounts. The blue arrows represent basic nutrient pathways such as plant uptake, biomass decomposition, and nutrient return to soil. In the agroecosystem fertilizer (e.g., manure, compost, chemical) is applied and nutrients are removed through plant harvest (red arrows). Greater potential leaching (e.g., nitrate), erosion (e.g., soil, phosphate), and CO_2 emissions in the agroecosystem are indicated via thick black arrows.

It is generally accepted that there are 17 essential elements required for plant growth. The lack of any one of these essential nutrients, listed in Table 1, can result in a severe limitation of crop yield — an example of the principle of limiting factors. Of the mineral elements, the primary macronutrients (N, P, and K) are needed in the greatest quantities from the soil and are the plant nutrients most likely to be in short supply in agricultural soils. Secondary macronutrients are needed in smaller quantities, are typically in sufficient quantities in soil, and therefore are not often limiting for crop growth. The micronutrients, or sometimes called trace nutrients, are needed in very small amounts and, if in excess, can be toxic to plants. Silicon (Si) and sodium (Na) are sometimes considered to be essential plant nutrients, but due to their ubiquitous presence in soils they are never in short supply.

Essential plant element		Symbol	Primary form
Non-Mineral Elements			
	Carbon	C	CO_2 (g)
	Hydrogen	H	H_2O (l), H^+
	Oxygen	O	H_2O (l), O_2(g)
Mineral Elements			
Primary Macronutrients	Nitrogen	N	NH_4^+, NO_3^-
	Phosphorus	P	HPO_4^{2-}, $H_2PO_4^-$
	Potassium	K	K^+
Secondary Macronutrients	Calcium	Ca	Ca^{2+}
	Magnesium	Mg	Mg^{2+}
	Sulfur	S	SO_4^{2-}
Micronutrients	Iron	Fe	Fe^{3+}, Fe^{2+}
	Manganese	Mn	Mn^{2+}
	Zinc	Zn	Zn^{2+}
	Copper	Cu	Cu^{2+}
	Boron	B	$B(OH)_3$
	Molybdenum	Mo	MoO_4^{2-}
	Chlorine	Cl	Cl^-
	Nickel	Ni	Ni^{2+}

Agriculture alters the natural cycling of nutrients in soil. Intensive cultivation and harvesting of crops for human or animal consumption can effectively mine the soil of plant nutrients. In order to maintain soil fertility for sufficient crop yields, soil amendments are typically required. Early humans soon learned to amend their fields with animal manure, charcoal, ash, and lime (CaCO3) to improve soil fertility. Today, farmers add numerous soil amendments to enhance soil fertility, including inorganic chemical fertilizers and organic sources of nutrients, such as manure or compost, often resulting in surplus quantities of primary macronutrients. The efficiency of fertilizer application and use by crops is not always optimized, and excess nutrients, especially N and P, can be transported via surface runoff or leaching from agricultural fields and pollute surface- and groundwater.

Soils for Agriculture

While soil is frequently referred to as the "fertile substrate", not all soils are suitable for growing crops. Ideal soils for agriculture are balanced in contributions from mineral components (sand: 0.05–2 mm, silt: 0.002–0.05 mm, clay: <0.002 mm), soil organic matter (SOM), air, and water. The balanced contributions of these components allow for water retention and drainage, oxygen in the root zone, nutrients to facilitate crop growth; and they provide physical support for plants. The distribution of these soil components in a particular soil is influenced by the five factors of soil formation: parent material, time, climate, organisms, and topography. Each one of these factors plays a direct and overlapping role in influencing the suitability of a soil for agriculture.

Inorganic Soil Components

Loamy-textured soils are commonly described as medium textured with functionally-equal contributions of sand, silt, and clay. These medium-textured soils are often considered ideal for agriculture as they are easily cultivated by farmers and can be highly productive for crop growth.

The mineral components of soil may exist as discrete particles, but are more commonly associated with one another in larger aggregates that provide structure to soil. These aggregates, or peds, play an important role in influencing the movement of water and air through soil. Sandy soils have large pore spaces and increase water drainage, but do not provide soils with many nutrients. Clay-rich soils, on the other hand, increase water holding capacity and provide many plant essential nutrients. A common measure of soil fertility is obtained by measuring the cation exchange capacity (CEC). The CEC is a measure of a soil's ability to exchange positive ions between the soil particles and solution surrounding these particles.

Due to their high surface area, clay particles can exert a large influence on various soil properties (e.g., CEC, structure, water-holding capacity), even when the percent clay content is low. Clay minerals are colloidal particles, having high surface area, with charged surfaces; permitting binding of many essential plant nutrients. The most prevalent clay-sized particles in soils fall into the class of layer-type aluminosilicates that commonly have permanent negative charge with a high CEC. Positively charged clay particles, which bind anions, include those which have pH dependent charge. The most common classes of these minerals in soils are the iron (Fe), aluminum (Al), and manganese (Mn) (hydr)oxides.

Soil Organic Matter (SOM)

Mollisol soil profile showing thick dark A horizon with high organic matter content.

SOM comprises the partial or well-decomposed residues of organic biomass present in soil. SOM gives topsoil its deep black colors and rich aromas that many home gardeners and farmers of grassland soils are familiar with. Surface soils are composed of approximately 1 to 6% organic matter, with SOM decreasing with depth. The 'Great Plains' of North America and the 'Bread Basket' of Europe are some of the world's most productive agricultural soils because they developed under grassland vegetation, whose root biomass and decomposition resulted in SOM accumulation. The thick dark upper layers in this soil reflect the high SOM content. The presence of SOM is crucial for fertile soil as it provides essential plant nutrients, beneficially influences soil structure, buffers soil pH, and improves water holding capacity and aeration. The presence of organic, ionizable functional groups (e.g., carboxyl, alcoholic/phenolic OH, enol, quinone, and amine) impart charge to SOM, contributing high CEC, and pH buffering capacity.

Soil pH

Often referred to as the master variable of soil, pH controls a wide range of physical, chemical, and biological processes and properties that affect soil fertility and plant growth. Soil pH, which reflects the acidity level in soil, significantly influences the availability of plant nutrients, microbial activity, and even the stability of soil aggregates. At low pH, essential plant macronutrients (i.e., N, P, K, Ca, Mg, and S) are less bioavailable than at higher pH values near 7, and certain micronutrients (i.e., Fe, Mn, Zn) tend to become more soluble and potentially toxic to plants at low pH values (5–6). Aluminum toxicity is also a common problem for crop growth at low pH (<5.5). Typically, soil pH values from 6 to 7.5 are optimal for plant growth; however, there are certain plants species that can tolerate — or even prefer — more acidic or basic conditions. Maintaining a narrow range in soil pH is beneficial to crop growth. SOM and clay minerals help to buffer soils to maintain a pH range optimal for plant growth. In instances where the pH is outside a desirable range, the soil pH can be altered through amendments such as lime to raise the pH. Ammonium sulfate, iron sulfate, or elemental sulfur can be added to soil to lower pH.

Soil Degradation and Crop Production

Soil forms from fresh parent material through various chemical and physical weathering processes and SOM is incorporated into soil through decomposition of plant residues and other biomass. Although these natural soil building processes regenerate the soil, the rate of soil formation is very slow. For this reason, soil should be considered a nonrenewable resource to be conserved with care for generations to come. The rate of soil formation is hard to determine and highly variable, based on the five factors of soil formation. Scientists have calculated that 0.025 to 0.125 mm of soil is produced each year from natural soil forming processes. Because of the time required to generate new soil, it is imperative that agricultural practices utilize best management practices (BMPs) to prevent soil erosion. The soil which is first eroded is typically the organic and nutrient enriched surface layer which is highly beneficial for plant growth. Thus, the primary on-site outcome is reduced crop yield as only the less fertile subsurface layers remain. Soil erosion also pollutes adjacent streams and waterways with sediment, nutrients, and agrochemicals creating serious off-site impacts.

Erosion is often accelerated by agricultural practices that leave the soil without adequate plant cover and therefore exposed to raindrop splash and surface runoff or wind. Throughout human history, soil erosion has affected the ability of societies to produce an adequate food supply.

Measurement type	Sample size n	Mean rate of soil erosion (mm/yr)	Net rate of soil formation (mm/yr)
Conventional Agriculture	448	1.54 (0.32)	-1.52 to -1.42
Conservation Agriculture	47	0.082 (0.022)	-0.057 to +0.043

Native Vegetation	65	0.013 (0.016)	+0.012 to +0.112
Geological	925	0.029 (0.029)	-4.00×10^{-3} to +0.096

Dallas, South Dakota: tractor and farm equipment buried by soil transported by wind.

Today, agricultural fields are not immune to the forces of nature (e.g., moving water, blowing wind, extremes of temperature) that caused soil erosion in the past. Figure below shows the severe effects of surface runoff and soil loss in the northwestern United States. Implementation of agricultural best management practices (BMPs), and through the practice of conservation agriculture, the rate of soil loss can be reduced to approximately equal the rate of soil formation, although often still greater than that in natural systems. In addition to soil erosion, intensive land use has resulted in deforestation, water shortages, and rapidly increasing desertification of vast areas of the globe, all of which threaten the sustainability of our agricultural systems.

Damage to agricultural field in Washington (United States) resulting from water erosion.

Sustainable Soil Management

It is evident that, in order to maintain and increase food production, efforts to prevent soil degradation must become a top priority of our global society. Current population models predict a global population of between 8 and 10 billion in the next 50 years and a two-fold increase in food demand.

If mismanagement of soil resources continues to diminish the fertility of the soil and the amount of productive arable land, then we will have lost a precious and essential pillar of sustainable agriculture. Sustainable agriculture is an approach to farming that focuses on production of food in a manner that can be maintained with minimal degradation of ecosystems and natural resources. This sustainable approach to agriculture strives to protect environmental resources, including soil, and provide economic profitability while maintaining social equity. The concept of sustainable agriculture is often misinterpreted to mean that chemical fertilizers and pesticides should never be used. This notion is incorrect, as sustainable agriculture should embrace those practices that provide the most beneficial services for agroecosystems and encourage long-term production of food supplies in a cultural context of the region. It cannot be overstressed that sustainable practices should not only consider crop production and profit, but must include land management strategies that reduce soil erosion and protect water resources. By embracing certain modern-day technologies, proven BMPs, and learning from the past, our society will be able to continue to conserve soil resources and produce food supplies sufficient to meet current and future population demands.

Soil Nutrients

Soil is a major source of nutrients for plant growth. Nutrients supplied by the soil are called mineral nutrients. The non-mineral nutrients such as carbon (C), hydrogen (H) and oxygen (O) come from air and water during photosynthesis. Soil mineral nutrients are separated into two groups the macro and micronutrients. The macro nutrients are further broken down into two groups the primary and the intermediate nutrients. The primary nutrients are required by plants in relatively large proportions. These are the most famous; the nitrogen (N), phosphorus (P) and potassium (K) commonly referred to as NPK. The intermediate nutrients are required by plants in medium quantities, these are calcium (Ca) magnesium (Mg) and sulphur (S).

The micronutrients are required in relatively small proportions. They include the iron (Fe), boron (B), manganese (Mn), copper (Cu), zinc (Zn), molybdenum (Mo), nickel (Ni) and chlorine (Cl). It is important to note that though the soil nutrients are separated into different groups (based on the quantity required by the plant), each nutrient is equally important. A shortage of any nutrient can limit the growth and yield of a plant. This is in accordance with Liebegs law of the minimum.

Forms by which Nutrient Elements are Taken up by Plants:

Element	Chemical symbol	Ionic forms Absorbed by plants
Macronutrients		
Nitrogen	N	NO_3^-, NH_4^+
Phosphorus	P	PO_4^{3-}, HPO_4^{2-}, $H_2PO_4^-$
Potassium	K	K^+
Magnesium	Mg	Mg^{2+}
Sulfur	S	SO_4^{2-}
Calcium	Ca	Ca^{2+}
Micronutrients		
Iron	Fe	Fe^{2+}, Fe^{3+}
Manganese	Mn	Mn^{2+}
Zinc	Zn	Zn^{2+}
Copper	Cu	Cu_2^+
Boron	B	BO_3^{2-}, $B_4O_7^{2-}$
Molybdenum	Mo	MoO_4^{2-}
Chlorine	Cl	Cl^-

Soil Nutrient Sources

Sources of soil nutrient include:

1. Organic matter decomposition,
2. Precipitation,
3. Biological nitrogen fixation,
4. Inorganic fertilizer application,
5. Weathering of soil rocks and minerals.

What affects the availability of nutrients in soil?

Factors that affect the availability of soil nutrients include leaching, soil erosion, soil pH, denitrification, volatilization, nitrogen immobilization and crop nutrient uptake.

Soil Erosion

Soil Erosion disrupts the soil structure, washes away organic matter in the soil and therefore reducing soil fertility. This often increases the need for additional and costly fertilizers to compensate for nutrient losses.

Some causes of soil erosion include:

1. Soil gradient: The steeper the gradient of a soil, the more vulnerable it is to soil erosion.
2. Soil erodibility: The more erodible a soil, the more sensitive it is to erosion. This is influenced by the soil characteristic and nature. A soil that has experienced previous erosion will have higher chances of erosion.
3. Vegetative cover: Soil vegetation can protect the soil from wind or water erosion by creating vegetative cover to the soil.

Leaching

This is the washing downward of nutrients in the soil below the root zone. Some of the factors influencing leaching include:

1. Mobility of nutrients: When there is sufficient water in the soil, nutrients in soil solution can be easily washed down beyond the root zone. An example is nitrogen present in the form of nitrate; a highly mobile negatively charged ion.
2. Soil texture: Leaching occurs in soils which have high water infiltration rates and low ability to hold nutrients. Examples of such are the sandy soil and clay soil.

Denitrification

This process results in the gaseous loss of nitrogen to the atmosphere. Denitrification occurs in warm and anaerobic (saturated) soils usually having high nitrate levels. During this process soil microbes break down nitrate to obtain oxygen for their respiration. The end product is nitrogen gas released to the atmosphere. Ways of avoiding loss of nitrogen through denitrification include proper timing of organic or inorganic fertilizer application; such that the soil receives it when it really needs it. Crop producers are advised to apply fertilizers in splits in order to match the crop demand for nitrogen with supply.

$NO_3^- \rightarrow$	$NO_2^- \rightarrow$	$NO \longrightarrow$	$N_2O \longrightarrow$	N_2
Nitrate	Nitrite	Nitric oxide	Nitrous oxide (gas)	Nitrogen gas

Stages of denitrification reaction.

Volatilization

This process also involves the gaseous loss of nitrogen to the atmosphere. When nitrogen fertilizers are applied in urea form (this could be inorganic fertilizers of animal manure), an enzyme urease catalyzes the reaction of urea with water resulting in ammonia gas released to the atmosphere. Volatilization occurs in warm and moist soil conditions. Volatilization also increases with high soil pH. To reduce nitrogen loss through this means, crop producers are encouraged to adapt methods of applying urea or ammonium based fertilizers below the soil surface.

Soil Fertility

Soil fertility refers to the ability of soil to sustain agricultural plant growth, i.e. to provide plant habitat and result in sustained and consistent yields of high quality. A fertile soil has the following properties:

- The ability to supply essential plant nutrients and water in adequate amounts and proportions for plant growth and reproduction; and
- The absence of toxic substances which may inhibit plant growth.

The following properties contribute to soil fertility in most situations:

- Sufficient soil depth for adequate root growth and water retention;
- Good internal drainage, allowing sufficient aeration for optimal root growth (although some plants, such as rice, tolerate waterlogging);
- Topsoil with sufficient soil organic matter for healthy soil structure and soil moisture retention;
- Soil pH in the range 5.5 to 7.0 (suitable for most plants but some prefer or tolerate more acid or alkaline conditions);
- Adequate concentrations of essential plant nutrients in plant-available forms;
- Presence of a range of microorganisms that support plant growth.

In lands used for agriculture and other human activities, maintenance of soil fertility typically requires the use of soil conservation practices. This is because soil erosion and other forms of soil degradation generally result in a decline in quality with respect to one or more of the aspects indicated.

Soil Fertilization

Bioavailable phosphorus is the element in soil that is most often lacking. Nitrogen and potassium are also needed in substantial amounts. For this reason these three elements are always identified on a commercial fertilizer analysis. For example, a 10-10-15 fertilizer has 10 percent nitrogen, 10 percent (P_2O_5) available phosphorus and 15 percent (K_2O) water-soluble potassium. Sulfur is the fourth element that may be identified in a commercial analysis—e.g. 21-0-0-24 which would contain 21% nitrogen and 24% sulfate.

Inorganic fertilizers are generally less expensive and have higher concentrations of nutrients than organic fertilizers. Also, since nitrogen, phosphorus and potassium generally must be in the inorganic forms to be taken up by plants, inorganic fertilizers are generally immediately bioavailable to plants without modification. However, some have criticized the use of inorganic fertilizers, claiming that the water-soluble nitrogen doesn't provide for the long-term needs of the plant and creates water pollution. Slow-release fertilizers may reduce leaching loss of nutrients and may make the nutrients that they provide available over a longer period of time.

Soil fertility is a complex process that involves the constant cycling of nutrients between organic and inorganic forms. As plant material and animal wastes are decomposed by micro-organisms, they release inorganic nutrients to the soil solution, a process referred to as mineralization. Those nutrients may then undergo further transformations which may be aided or enabled by soil micro-organisms. Like plants, many micro-organisms require or preferentially use inorganic forms of nitrogen, phosphorus or potassium and will compete with plants for these nutrients, tying up the nutrients in microbial biomass, a process often called immobilization. The balance between immobilization and mineralization processes depends on the balance and availability of major nutrients and organic carbon to soil microorganisms. Natural processes such as lightning strikes may fix atmospheric nitrogen by converting it to (NO_2). Denitrification may occur under anaerobic conditions (flooding) in the presence of denitrifying bacteria. Nutrient cations, including potassium and many micronutrients, are held in relatively strong bonds with the negatively charged portions of the soil in a process known as cation exchange.

In 2008 the cost of phosphorus as fertilizer more than doubled, while the price of rock phosphate as base commodity rose eight-fold. Recently the term peak phosphorus has been coined, due to the limited occurrence of rock phosphate in the world.

Light and CO_2 limitations

Photosynthesis is the process whereby plants use light energy to drive chemical reactions which convert CO_2 into sugars. As such, all plants require access to both light and carbon dioxide to produce energy, grow and reproduce.

While typically limited by nitrogen, phosphorus and potassium, low levels of carbon dioxide can also act as a limiting factor on plant growth. Peer-reviewed and published scientific studies have shown that increasing CO_2 is highly effective at promoting plant growth up to levels over 300 ppm. Further increases in CO_2 can, to a very small degree, continue to increase net photosynthetic output.

Soil Depletion

Soil depletion occurs when the components which contribute to fertility are removed and not replaced, and the conditions which support soil's fertility are not maintained. This leads to poor crop yields. In agriculture, depletion can be due to excessively intense cultivation and inadequate soil management.

Soil fertility can be severely challenged when land use changes rapidly. For example, in Colonial New England, colonists made a number of decisions that depleted the soils, including: allowing herd animals to wander freely, not replenishing soils with manure, and a sequence of events that led to erosion. William Cronon wrote that "the long-term effect was to put those soils in jeopardy. The removal of the forest, the increase in destructive floods, the soil compaction and close-cropping wrought by grazing animals, plowing all served to increase erosion."

Karl Marx wrote of the role of capitalism in soil depletion.

All progress in capitalistic agriculture is a progress in the art, not only of robbing the labourer, but of robbing the soil; all progress in increasing the fertility of the soil for a given time, is a progress towards ruining the lasting sources of that fertility. The more a country starts its development on the foundation of modern industry, like the United States, for example, the more rapid is this process of destruction. Capitalist production, therefore, develops technology, and the combining together of various processes into a social whole, only by sapping the original sources of all wealth — the soil and the labourer.

One of the most widespread occurrences of soil depletion as of 2008 is in tropical zones where nutrient content of soils is low. The combined effects of growing population densities, large-scale industrial logging, slash-and-burn agriculture and ranching, and other factors, have in some places depleted soils through rapid and almost total nutrient removal.

The depletion of soil has affected the state of plant life and crops in agriculture in many countries. In the middle east for example, many countries in that find it difficult to grow produce because of droughts, lack of soil, and lack of irrigation. "The Middle East" has three countries that indicate a decline in crop production. the highest rates of productivity decline are found in hilly and dryland areas. Many countries in Africa also undergo a depletion of fertile soil. In regions of dry climate like Sudan and the countries that make up the Sahara Desert, droughts and soil degradation is common. Cash crops such as teas, maize, and beans that require a variety of nutrients in order to grow healthy. Soil fertility has decline in the farming regions of Africa and the use of artificial and natural fertilizers has been used to regain the nutrients of ground soil.

Topsoil depletion occurs when the nutrient-rich organic topsoil, which takes hundreds to thousands of years to build up under natural conditions, is eroded or depleted of its original organic material. Historically, many past civilizations' collapses can be attributed to the depletion of the topsoil. Since the beginning of agricultural production in the Great Plains of North America in the 1880s, about one-half of its topsoil has disappeared.

Depletion may occur through a variety of other effects, including overtillage (which damages soil structure), underuse of nutrient inputs which leads to mining of the soil nutrient bank, and salinization of soil.

Irrigation Water Effects

The quality of irrigation water is very important to maintain soil fertility and tilth, and for using more soil depth by the plants. When soil is irrigated with high alkaline water, unwanted sodium salts build up in the soil which would make soil draining capacity very poor. So plant roots can not penetrate deep into the soil for optimum growth in Alkali soils. When soil is irrigated with low pH / acidic water, the useful salts (Ca, Mg, K, P, S, etc.) are removed by draining water from the acidic soil and in addition unwanted aluminium and manganese salts to the plants are dissolved from the soil impeding plant growth. When soil is irrigated with high salinity water or sufficient water is not draining out from the irrigated soil, the soil would convert into saline soil or lose its fertility. Saline water enhance the turgor pressure or osmotic pressure requirement which impedes the off take of water and nutrients by the plant roots.

Top soil loss takes place in alkali soils due to erosion by rain water surface flows or drainage as they form colloids (fine mud) in contact with water. Plants absorb water-soluble inorganic salts only from the soil for their growth. Soil as such does not lose fertility just by growing crops but it lose its fertility due to accumulation of unwanted and depletion of wanted inorganic salts from the soil by improper irrigation and acid rain water (quantity and quality of water). The fertility of many soils which are not suitable for plant growth can be enhanced many times gradually by providing adequate irrigation water of suitable quality and good drainage from the soil.

Soil Texture

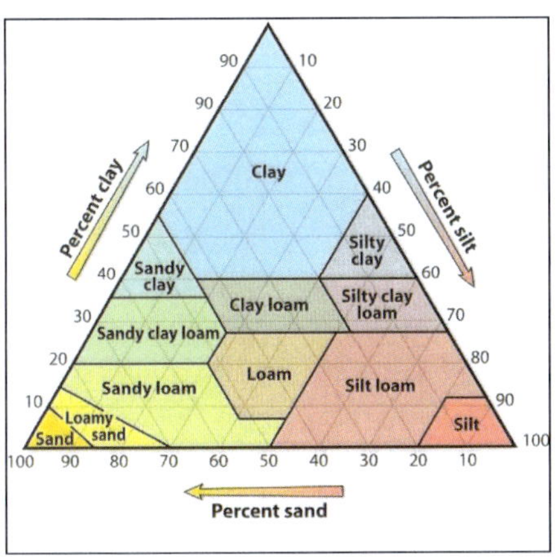

Textural triangle used in determining soil textural class.

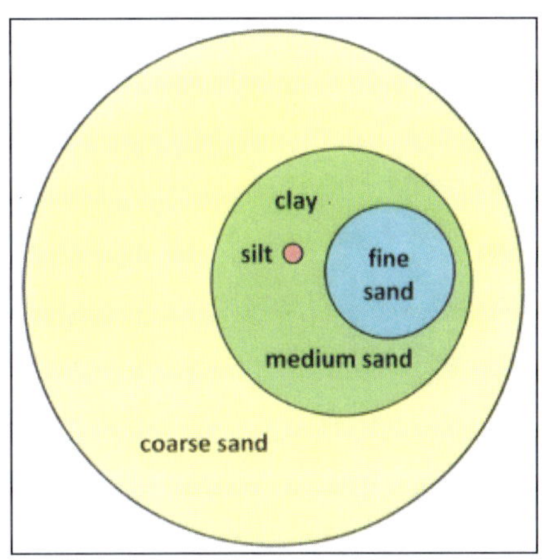

Relative size of soil particles.

Soil texture refers to a mixture of variously sized mineral particles (right), which determine a soil's textural class. The textural class is defined by the relative amounts of sand (0.05 to 2 mm particle size), silt (0.002 to 0.05 mm), and clay (less than 0.002 mm) as seen in the textural triangle above. Particles that are larger than 2 mm are called coarse fragments (pebbles, cobbles, stones, and boulders), and are not considered in the textural class, although they may help define a soil type. Organic matter is also not considered in the determination of soil texture, although it is very important for soil functioning. A soil's textural class—such as a clay, clay loam, loam, sandy loam, or sand—is perhaps its most fundamental inherent characteristic. It affects many of the important physical, biological, and chemical processes in a soil, but is not easily altered by management, and changes little over time. Thus, while texture is not a soil health indicator per se, it informs the interpretation of most soil health indicators.

Relating Texture to Soil Function

Texture affects many important soil processes due to the total amount of pore space and how varied pore space is within aggregates. Soils with higher clay contents (finer soils) generally have higher ability to retain nutrients (more cation exchange capacity, or CEC) and can accumulate, or sequester, more organic matter. The size distribution of the particles strongly influences the size of the pore spaces between the particles, the formation and stabilization of soil aggregates, and the spaces between these aggregates. Aggregates and interaggregate spaces are as important as the sizes of the particles themselves, because the relative quantities of variously sized pores—large, medium, small, and very small—govern the important processes of water and air movement. These in turn affect processes like water infiltration, permeability, water storage, aeration, nutrient leaching, and denitrification. In addition, soil organisms and plant roots live and function in the pores. When the soil loses porosity (generally due to management), roots cannot grow as well, and many organisms have more difficulty surviving. Most pores in a clay are small (generally less than 0.002 mm), whereas most pores in a sand are large (but generally still smaller than 2 mm).

On the one extreme of the texture and aggregation spectrum, we see that beach sands have large particles (in relative terms) and very poor aggregation due to a lack of organic matter or clay to help bind the sand grains. A good loam or clay soil, on the other hand, has smaller particles, but they tend to be aggregated into crumbs that have larger pores between them and small pores within. Although soil texture doesn't generally change over time, the total amount of pore space and the relative amount of variously sized pores are strongly affected by management practices.

Using texture in developing scoring functions Soil texture contributes to inherent soil quality, and is virtually unchangeable through soil management. However, most interpretations of soil health indicator scores are dependent on interactions with soil texture. Therefore texture is not given a soil health score. For example, given the same management, coarse textured soils generally have lower organic matter (OM) levels than fine-textured soils because they lack the ability to stabilize OM. Measured OM contents, along with other indicators, are scored relative to an appropriate distribution for soils of a particular textural grouping, to account for this type of difference.

In the soil health assessment scoring process, we distinguish between coarse-textured (sand, loamy sand, sandy loam), medium-textured (loam, silt loam, silt, sandy clay loam) and fine-textured (clay loam, silty clay loam, sandy clay, silty clay, and clay) soils.

Basic Protocol

1. Air dry and sieve soil sample past 2 mm.
2. Approximately 14g (+/- 0.1g) of sieved soil is added to a 50ml centrifuge tube containing 42ml of a dispersant solution (3% sodium hexametaphosphate, a detergent).
3. Shake vigorously on reciprocating shaker for 2 hours to fully disperse soil into suspension.
4. Wash entire contents of centrifuge tube onto a sieve assembly. Sieve assembly consists of 0.053mm sieve on top of a funnel above a 1L beaker. Rinse all material through the sieve. Sand captured on top of the sieve is washed into a tared metal can and set aside.
5. Silt and clay particles collected in the 1L beaker are re-suspended by stirring and allowed to settle for 2 hours. The clay in suspension is then decanted. The settled silt is washed into a second tared can. Both tared cans (one containing the sand fraction and the other the silt fraction) are dried at 105O C to constant weight before recording dry weight.
6. Calculate percent sand, silt clay from:
   - Sand (%) = dry wt sand (g)/dry wt soil (g) added to centrifuge tube.
   - Silt (%) = dry wt silt (g)/dry wt soil (g) added to centrifuge tube.
   - Clay (%) = 100% - Sand (%) - Silt (%).

Classes of Soil Texture

Texture names are given to soils based upon the relative proportion of each of the three soil separates sand, silt and clay. Soil that are preponderantly clay, are called clay (textural class), those with high silt content are silt (textural class) those with high sand percentage are sand (textural class). Three broad and fundamental groups of soil texture classes are recognised: sands, loams and clays.

1. Sands: The sand group includes all soils of which the sand separates make up 70 per cent or more of the material by weight. Two specific classes are recognised sand and loamy sand.
2. Loams: A loamy soil containing many sub-divisions does not exhibit the dominant physical properties of any of these three soils separates sand, silt and clay. An ideal loam soil may be defined as a mixture of sand, silt and clay particles which exhibits light and heavy properties in about equal proportions.

   Note that loam does not contain equal percentages of sand, silt and clay. It does, however, exhibit approximately equal properties of sand, silt and clay.
3. Clay: A clay soil must carry at least 35 per cent of the clay separate and in most cases not less than 40 per cent. For an example, sandy clay soils contain more sand than clay.

Similarly silty clay soils contain more silt than that of the clay. Based on these three broad and fundamental groups, the different textural class names developed by U.S. Department of Agriculture and U.S. Bureau of soils are presented in table.

Texual class names developed by U.S Department of agriculture:

Common name	**Texture**	**Basic soil textural class names**
Sandy soils	Coarse	Sand Loamy sand
Loamy soils	Moderately coarse	Sandy loam Fine sandy loam
	Medium	Very fine sandy loam Loam Silt loam Silt
	Moderately fine	Sandy clay loam Silty clay loam Clay loam
Clayey soils	Fine	Sandy clay Silty clay Clay

Textural groups on the basis o percentage of sand, silt and clay Separates:

Textural group	***Sand***	***Silt***	***Clay***
1. Sand	80-100	0-20	0-20
2. Sandy loam	50-80	0-50	0-20
3. Loam	30-50	30-50	0-20
4. Silt loam	0-50	50-100	0-20
5. Sandy clay loam	50-80	0-30	20-30
6. Silty clay loam	0-30	50-80	20-30
7. Clay loam	20-50	20-50	20-30
8. Sandy clay	50-70	0-20	30-50
9. Silty clay	0-20	50-70	30-50
10. Clay	0-50	0-50	30-100

Determination of Soil Texture

1. Feel Method:

   In the field, texture is commonly determined by the sense of feel. The soil is rubbed between thumb and fingers under wet conditions. Sands feel gritty and its particles can be easily seen. The silt when dry feels like flour and talcum powder and is slightly plastic when wet. Clayey particles feel very plastic and exhibit stickiness when wet and are hard under dry conditions.

2. Laboratory Method:

   A more accurate and fundamental method has been devised by the U.S. Department of Agriculture for the naming of soils based on a mechanical analysis.

References

- Soil-the-foundation-of-agriculture: nature.com, Retrieved 15 June, 2019
- Soil-nutrients: permaculturenews.org, Retrieved 5 May, 2019
- F. Stuart Chapin III; Pamela A. Matson; Harold A. Moon (2002). Principles of Terrestrial Ecosystem Ecology. Springer. ISBN 0387954392
- Kötke, William H. (1993). The Final Empire: The Collapse of Civilization and the Seed of the Future. Arrow Point Press. ISBN 0963378457

Chapter 5

Pests and their Management

A pest is any living organism which is harmful to plants, animals, humans, crops, livestock and forestry. Potato bug, locust and corn rootworms are some of the pests which affect agriculture. The different types of pests and how to manage them has been thoroughly discussed in this chapter.

Plant Pests and Diseases

Transboundary plant pests and diseases affect food crops, causing significant losses to farmers and threatening food security.

The spread of transboundary plant pests and diseases has increased dramatically in recent years. Globalization, trade and climate change, as well as reduced resilience in production systems due to decades of agricultural intensification, have all played a part.

Transboundary plant pests and diseases can easily spread to several countries and reach epidemic proportions. Outbreaks and upsurges can cause huge losses to crops and pastures, threatening the livelihoods of vulnerable farmers and the food and nutrition security of millions at a time.

Locusts, armyworm, fruit flies, banana diseases, cassava diseases and wheat rusts are among the most destructive transboundary plant pests and diseases. Plant pests and diseases spread in three principal ways:

- Trade or other human-migrated movement.
- Environmental forces-weather and windborne.
- Insect or other vector-borne-pathogens.

Cassava Virus Diseases

Cassava Mosaic and Brown Streak virus diseases continue to affect the main food crop – cassava – throughout the Great Lakes region of Eastern and Southern Africa. In Africa, an estimated 70 million people are dependent on cassava as a primary source of food contributing over 500 kcal per day per person.

Cassava is produced mostly by smallholders on marginal and sub-marginal lands in the humid and semi-humid tropics. It is efficient in carbohydrate production, adapted to a wide range of environments and tolerant to drought and acidic soils.

Desert Locust

The Desert Locust migrates in swarms across continents and is a potential threat to the livelihood of one-tenth of the world's population. This pest is a serious menace to agricultural production in Africa, the Near East and Southwest Asia. A locust can eat its own weight (about 2 grams) in plants every day. That means one million locusts can eat about one tonne of food each day, and the largest swarms can consume over 100 000 tonnes each day, or enough to feed tens of thousands of people for one year.

Wheat Rusts

Wheat rust diseases with continuous evolution of new pathotypes and airborne nature pose a serious threat to wheat production worldwide. Their impact is more pronounced across the major wheat growing regions including East Africa, North Africa, Middle East and Asia. It is estimated that 37% of world's wheat is under risk of potential epidemics of yellow, stem or leaf rust diseases.

Some Common Agricultural Pests

Locust

Locust is any of a group of insects (order Orthoptera) that are distributed worldwide, the common name of which generally refers to the group of short-horned grasshoppers that often increase greatly in numbers and migrate long distances in destructive swarms. In Europe the term locust denotes large acridids, whereas smaller species are called grasshoppers. In North America the names locust and grasshopper are used for any acridid. Cicadas (order Homoptera) also may be called locusts, the 17-year "locust" being the 17-year periodic cicada. The grouse (or pygmy) locust is a member of the family Tetrigidae.

A phase theory has been developed to account for the sporadic appearance and disappearance of locust swarms. According to the theory, a plague species has two phases: one solitary and the other gregarious. The phases can be distinguished by differences in coloration, form, physiology, and behaviour. A solitary-phase nymph, for example, adjusts its coloration to match that of its surroundings, does not collect in groups, has low metabolic and oxygen-intake rates, and is sluggish. A gregarious-phase nymph, on the other hand, has black and yellow or orange coloration in a fixed pattern, gathers in large groups, has high metabolic and oxygen intake rates, and is active and nervous. Adult locusts differ more in form than in colour. The solitary phase has shorter wings, longer legs, and a narrower pronotum, or dorsal sclerite (with higher crest and larger head), than the gregarious phase. The adult of the gregarious phase has a more saddle-shaped pronotum, broader shoulders, and longer wings.

When a nymph of a solitary-phase locust matures in the presence of many other locusts, it undergoes a physiological change and produces offspring of the gregarious type. If crowding is sufficiently dense and of long enough duration, the majority of a local population will shift to the gregarious migratory phase. The young of a gregarious-phase locust, on the other hand, will produce offspring that revert to the solitary phase if it matures in isolation. The solitary phase is the normal state of the species, the gregarious phase being a physiological response to violent fluctuations in the environment. Migratory swarms do not form in regions favourable for the growth of a species. Instead, they form in marginal regions in which suitable habitats are scarce. A succession of favourable seasons enables a population to expand in numbers so that individuals are forced into marginal areas. When unfavourable environmental conditions occur in the marginal regions, individuals are forced to return to smaller, permanently habitable areas, resulting in crowding and triggering the physiological shift to the gregarious form.

A gregarious-phase locust is restless and irritable, and it flies spontaneously on warm dry days, when its body temperature is high. The muscular activity of flight further raises its temperature. A swarm ceases flying only when environmental conditions change—e.g., rain falls, temperature decreases, or darkness occurs. In 1869 desert locust swarms reached England, probably from West Africa, and a flight across the Red Sea in 1889 was estimated to be about 5,000 square km (2,000 square miles) in size. The long-distance dispersal of these swarms is usually associated with either frontal winds of storm systems or high-level jet-stream winds. The acridids typically fly almost straight up into these fast-moving winds and then are carried with the winds until they slow to the point where gravity overcomes wind speed, causing them to drop from the sky.

The range of the migratory locust (Locusta migratoria) is wider than that of any other acridid. It is found in grasslands throughout Africa, most of Eurasia south of the taiga, the East Indies, tropical Australia, and New Zealand. The desert locust (Schistocerca gregaria) inhabits dry grasslands and deserts from Africa to the Punjab and can fly upward to about 1,500 metres (5,000 feet) in huge towers of individuals. The smaller Italian and Moroccan locusts (Calliptamus italicus and Dociostaurus maroccanus) cause extensive plant damage in the Mediterranean area, with D. maroccanus found as far east as Turkestan. In South Africa the brown and red locusts (Locustana pardalina and Nomadacris septemfasciata) are extremely destructive. In Central and South America the chief migratory species is the South American locust (Schistocerca paranensis). The nonmigratory S. americana (found in the United States) may be a solitary phase of this genus. The Rocky Mountain locust and the migratory grasshopper (Melanoplus spretus and M. sanguinipes, respectively) destroyed many prairie farms in Canada and the United States in the 1870s. Many other species occasionally increase sufficiently in numbers to be called plagues.

Once developed, a locust plague is almost impossible to stop or control. Control measures include destroying egg masses laid by invading swarms, digging trenches to trap nymphs, using hopperdozers (wheeled screens that cause locusts to fall into troughs containing water and kerosene), using insecticidal baits, and applying insecticides to both swarms and breeding grounds from aircraft.

Impact on Agriculture

Locust swarms devastate crops and cause major agricultural damage and attendant human misery—famine and starvation. They occur in many parts of the world, but today locusts are most destructive in sustenance farming regions of Africa.

The desert locust (Schistocerca gregaria) is notorious. Found in Africa, the Middle East, and Asia, they inhabit some 60 countries and can cover one-fifth of Earth's land surface. Desert locust plagues may threaten the economic livelihood of one-tenth of the world's humans.

A desert locust swarm can be 460 square miles in size and pack between 40 and 80 million locusts into less than half a square mile.

Each locust can eat its weight in plants each day, so a swarm of such size would eat 423 million pounds of plants every day.

Distance Traveled

Like the individual animals within them, locust swarms are typically in motion and can cover vast distances. In 1954, a swarm flew from northwest Africa to Great Britain. In 1988, another made the lengthy trek from West Africa to the Caribbean.

Japanese Beetle

Japanese beetles are 3/8" (8-11 mm) long and ¼" (5-7 mm) wide, brilliant metallic green insects with copper-brown wings whose hard body makes them unpalatable to many predators, including birds. To distinguish them from other metallic green or copper-colored beetles, the diagnostic sign is a row of 5 small tufts of white hairs under the wing covers on each side and a 6th pair at the tip of the abdomen. The larvae, called grubs, are grayish-white with a dark brown head. They are C-shaped when disturbed. The 1st instar (stage of an insect or arthropod between molts) is 1/16" (1.5 mm) long and the 3rd instar is 1¼" (32 mm) long. They are found in the soil where they feed on the tender roots of vegetables, lawn grasses, and other plants.

Life Cycle

Japanese beetles overwinter as a partially grown grub in the soil below the frost line. The grubs resume feeding on grass roots in the spring, and then pupate near the soil surface. Adults emerge between May and July, depending on their geographic location. Adults live 30-45 days and feed, usually in groups, first on low-growing plants and then on tree leaves, working from the top of the plants downward. After feeding and mating, each female lays 1-5 eggs at a time before again feeding and mating until a total of 40-60 eggs are laid 2-4" under the soil surface in grassy areas. Most are laid by mid-August. The eggs hatch after 8-14 days, and the young grubs feed on fine grass roots until cold weather drives them below the frost line. Most are in the 3rd instar by this time. In spring the grubs come to the surface, continuing their development and forming a pupa in an earthen cell 1-3" deep in the soil.

Japanese beetle adults are active for about 6 weeks in the summer. Since adult beetle damage is primarily aesthetic, control is not essential for survival of ornamental plants. Because they are strong fliers and frequently move about, by season's end, adults are capable of having traveled many miles from where they lived as grubs. As a result, local beetle control does not insure less grub damage to lawns. Likewise, local grub control will not necessarily prevent adult damage to garden plants.

Integrated Pest Management Strategies:

- Hand-pick adult beetles. If only a few adults are present, shake plants early in the morning (~ 7 a.m.) when they are sluggish. They should be collected and dropped into a container of soapy water. Any beetles or damaged leaves will attract more insects and should be removed.
- Pheromone traps. Research has found that pheromone traps attract many more beetles than they catch and probably do more harm than good to plants in the beetles' flight path and near the traps. Their use is not recommended.

- Use insecticides if necessary. To control adults, one of two very safe pesticide, pyrethrum or Neem, can be applied in two applications, 3 to 4 days apart, to control the problem. If something stronger is needed, carbaryl (Sevin) may be used (every 5-10 days during heavy infestation) with monitoring for mites or aphids, in which case acephate (Orthene; more toxic) or malathion should be used. Other insecticides include Turcam, Closure (bendiocarb), permethrin, and synthetic pyrethroids. With any insecticide, efficacy will be decreased if there is heavy rainfall shortly after application of the chemical.

  To control grubs present in damaging numbers, newer chemicals, Merit (imidacloprid) and MACH2 (halofenozide), applied in June and July 20 days before anticipated Japanese beetle adult egg-laying activity, have enough residual to kill new grubs coming to the soil surface to feed in late July through August. It is important to apply evenly over all the ground and to water in well. These chemicals will also be effective on other grub species when applied at the appropriate times.

- Biological controls. The hard body of the beetle makes them unpalatable to many predators, such as birds. However, toads, moles, shrews, and skunks are known to feed on the beetles.

  To control grubs, bacterial milky spore disease (Paenibacillus popillae, formerly known as Bacillus popillae) can be applied as a dust for Japanese beetle grub control and must be ingested by the grub to be effective. Once inside the digestive tract; spores reproduce within the grub, eventually turning it an opaque milky white before killing it. Spores then disperse into the surrounding soil and can persist for many years, but do not spread until live hosts are present. The more grubs, the faster it spreads. Milky spore infects only Japanese beetle grubs and has no effect on beneficial organisms. Its efficacy is questionable and the spore count has to build up for 2-3 years, during which time no insecticides may be used. More research needs to be done.

  Other biological controls are parasitic nematodes that need to be applied when the grubs are small, with irrigation before and after application. Species of Heterorhabditis are said to be more effective than strains of Steinernema carpocapsae.

- Use trap crops. When practical, crops which are highly favored by Japanese beetles can be used to draw the adults away from other crops. Trap crops include: borage, white geraniums, grape vines, evening primroses, and zinnias. The beetles can be collected or killed on the trap crops. This method provides early detection and can limit the amount of spray needed to control the insects.

- Select resistant plants. Among trees and shrubs are: ash, dogwood, American elder, shagbark hickory, holly, red and silver maple, oaks, pear, white poplar, sweet gum, and tulip tree; euonymus and common lilac.

Corn Rootworms

Corn rootworms are important insect pest of corn in the Midwest. Two species of rootworms that may cause severe damage to corn as both larvae and adults, the western and northern corn rootworms. Southern corn rootworm adults may damage corn leaves, however, because they cannot overwinter in most areas of the Midwest, southern corn rootworm larvae do not present a major threat to corn in this region.

Both sexes of western corn rootworm adults are yellow to green in color with a black stripe along the sides of their wing covers and are about 5/16 inch (7.5 mm) long. However, male wing covers are often nearly entirely black or at least darker in pigmentation than that of the female, which usually appears as more regular stripes. Slight variations in these color patterns may occur. Female western corn rootworm adults have larger abdomens than the males.

The adult of the northern corn rootworm is a tan to pale green beetle about 1/4 inch (6 mm) long. Newly emerged beetles are usually cream or light brown in color, but gradually turn green with age. No marked differences in coloration exist between sexes, but female beetles with their longer and larger abdomens are typically larger than the males.

Southern corn rootworm.

The adult of the southern corn rootworm, also known as the spotted cucumber beetle, is about 3/8 inch (9 mm) long, yellow to green in color, with 11 conspicuous black spots on its back. Because adults and larvae of this species are considered of minor economic importance in the Midwest, the following discussion on damage, sampling, and management guidelines will not include the southern corn rootworm.

It is extremely difficult to distinguish between larvae of the western and northern species. Rootworm larvae are white and slender, about 1/2 inch (13 mm) long when fully grown, have brown heads, and a dark plate on the top side of their "tails."

Larva feeding on corn root.

Western and northern corn rootworm have only one generation per year. Eggs of both species are deposited in the soil by female beetles from mid-summer until autumn. The eggs overwinter and begin hatching from late May to early June in most areas of the Midwest. The newly hatched larvae seek out and feed on corn roots. The larvae pass through three stages, or instars, before pupating in the soil. Primary larval damage is caused by the later instars, typically from mid June through mid July.

By late June to early July, adult beetles begin to emerge from corn fields. The males emerge first, the females following in about 5 to 7 days. Mating takes place a short time thereafter. Within 2 weeks of emergence, females begin to lay eggs, usually toward the end of July, with peak egg laying occurring in early to mid August. The beetles in corn fields feed primarily on pollen, green silks, or leaves.

Several factors influence the development of a rootworm population in a particular field, and these should be considered when scouting. Rootworm beetles can readily move between fields and may cause damage in locations other than where they emerged. A late maturing corn field can attract large numbers of beetles since neighboring corn may have stopped producing pollen, the beetles' preferred food. Beetles may also move into corn and/or soybean fields which have an abundance of pollen-producing weeds, such as volunteer corn, ragweed, or foxtails. Some areas of the eastern Midwest have observed a steady and substantial increase of economic larval damage in first-year corn (corn following another crop). A behavioral variant of the western corn rootworm, that appears to have lost its fidelity for corn, can migrate to neighboring crops such as soybean to feed on pollen, flowers, and foliage, then lay eggs.

Northern corn rootworm.

Where northern corn rootworm is the dominant species, a trait referred to as "extended diapause" can affect the survivability of the insect. Extended diapause describes a condition whereby a portion of northern corn rootworm eggs are capable of remaining dormant in the soil through two winters and one growing season before hatching in the second season. First-year corn in areas of Minnesota, Iowa, South Dakota, and Wisconsin have been damaged by rootworms exhibiting this extended diapause trait. The western corn rootworm does not show this trait.

Soil texture is an important factor in regard to larval survival. Newly hatched rootworm larvae must move through the soil in search of corn roots and, as they move, they come in contact with soil particles. These particles, if coarse and abrasive, may scratch the cuticle of the larvae, resulting in death by desiccation. As a result, rootworms may be less of a problem in sandy soils. However, the effect of sandy soils can be modified by soil moisture. An irrigated or otherwise moist sandy soil will not affect rootworm populations as will the same soil under dry conditions. In addition, muck soils have shown a lower incidence of rootworm larval feeding damage and may provide some protection against this insect.

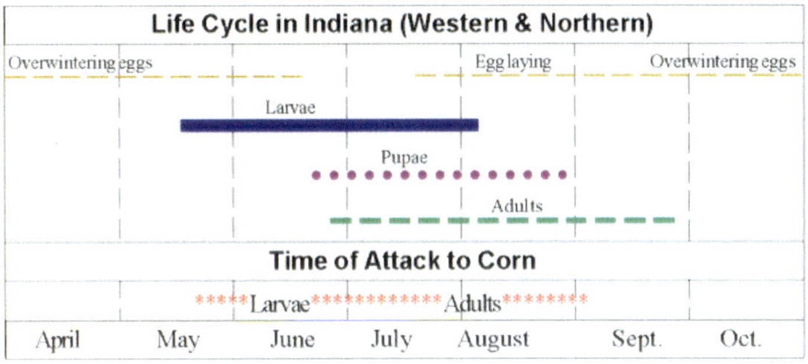

Damage

Western corn rootworm feeding.

Both corn rootworm larvae and adults may damage corn plants. Newly hatched larvae feed primarily on root hairs and outer root tissue. As larvae grow and their food requirements increase, they burrow into the roots to feed. Larval damage is usually most severe after the secondary root system is well established and brace roots are developing. Root tips will appear brown and are often tunneled into and chewed back to the base of the plant. Larvae may be found tunneling into larger roots and occasionally in the plant crown.

Varying degrees of poor pollination from silk feeding.

By reducing water and nutrient uptake of plants, larval root pruning places severe physiological stress on corn. Yield reductions may result, especially in corn also suffering from moisture, compaction, or fertility stress. Roots damaged by rootworms may be weakened to the extent that the plants lodge or grow in a curved "sled-runner" or "gooseneck" shape. These lodged and misshapen plants often pollinate poorly and are difficult to harvest, contributing to yield losses. Also, rootworm damage sites are often pathways for infection by pathogens. Thus, rootworm damage may be compounded by root and/or stalk rots.

High adult rootworm populations may interfere with corn plant pollination by severely clipping silks during pollen shed. This can result in poorly filled ears.

Corn rootworm damage.

Western corn rootworm beetles can also cause damage by feeding on the green tissue of the leaves. Leaves may be partially or totally stripped by adults, resulting in a parchment-like appearance to the damaged area. Unless these beetles are unusually abundant, only scattered plants are severely damaged. However, such foliar damage can be indicative of a large adult population which may be a serious threat to pollination and next season's corn roots.

Though considered of minor importance in other crops such as soybean and alfalfa, the variant of the western corn rootworm can feed on pollen, flowers, and foliage in these other crops. The presence of beetles and the extent of damage to crops such as soybean may indicate the potential of economic damage to next year's corn roots, if rotating to corn.

Sampling Method

Larvae in Corn - It will not be necessary to survey all corn fields for corn rootworm larvae. Sample those corn fields with suspected insecticide failures or fields in which a high adult rootworm population was observed, the previous July-August. Also sample corn fields that were in other crops like soybean the previous season, if beetles were observed or suspected as being present in those crops the previous year.

Randomly select 1 plant in each of at least 10 areas of a field. Using a spade or shovel, cut a 7 inch (18 cm) cube of soil around the base of each plant, making certain that the blade of the tool enters the ground vertically to avoid cutting roots. Lift the plant and soil out of the ground and place them on a small piece of dark canvas or plastic. Slowly break the soil away from the roots and carefully examine the soil and roots for larvae. The dark background will make it easier to find any of the small white rootworms. The soil and root sample can also be washed in a pail of water to extract the rootworms. The rootworms will float to the top and can be counted. The addition of salt to the water will float the worms to the top more easily.

Record the number of larvae found on each plant examined, repeating the procedure for each of the plants sampled. Determine the average number of rootworm larvae per plant by dividing the total number of larvae found in the field by the number of plants examined.

Adults in Corn - Surveys for rootworm beetles should begin shortly before corn silking commences. Several factors must be considered when planning adult rootworm scouting activities. Rootworm beetles are extremely active and can be readily observed feeding, mating, and flying about an infested field. However, such activity varies considerably with the time of day, and such behavior patterns can greatly influence field counts. Beetles are most active in the morning and late afternoon, limiting their activity during the normally higher temperatures of mid-day. Thus, to get the best indication of true population levels, sample in the morning or late afternoon.

Also, corn hybrids, even of the same maturity, vary to some degree in their time of pollination. Therefore, if a particular field has been planted to more than one hybrid (not blended together), each hybrid must be scouted separately.

To determine if rootworm adult sampling is necessary in a particular field once silking has or is about to begin, quickly walk across the field. If no beetles are observed in the ear zone area or on the ears, no further scouting will be necessary at this time. However, return to the field in 2 to 3 days to again check for beetles. Even if no beetles are found, you should check a field twice weekly until the silks have been pollinated and then once weekly until early September for egg laying activity.

1. Determining Potential for Silk Damage and Interference With Pollination:

   Randomly select and check 5 plants in at least 5 areas (25 total plants) of a field for silk clipping. If silk clipping is noted, determine (in inches) what length of the silk remains on each plant sampled. Determine the average silk length per plant by totaling the length for all plants sampled and dividing by 25.

   Record the number of beetles noted for each plant and determine the field average. Also, estimate and record the percent of pollination that has taken place. This can be done by carefully unwrapping the husk leaves from an ear and then gently shaking the ear. The silks from fertilized ovules will readily drop off the ear. As well, note if pollen is being shed by inspecting the tassels of each plant sampled. Remember that weather can greatly affect whether pollen in being shed. Record the results.

2. Determining Potential for Larval Problem in Next Year's Corn:

   Two randomly selected plants in each of at least 20 areas of the field should be checked. As you carefully approach each plant, watch for and record any rootworm beetles that drop from the plant or fly away. Grasp the silks in one hand to keep any beetles from leaving them while you inspect the remainder of the plant. After sampling the whole plant, slowly open your hand and count the number of beetles on or within the silks. Examine the silks thoroughly since beetles may be deep in the ear tip area. If there is a second ear on the plant, look through its silk. Repeat the procedure for the second plant that is within 10 feet of the first plant. After sampling all plants, determine the average number of beetles per plant. A face shield, or goggles, will protect your eyes from leaves and pollen while scouting corn fields for beetles.

Colorado Potato Beetle

The Colorado potato beetle (*Leptinotarsa decemlineata*), also known as the Colorado beetle, the ten-striped spearman, the ten-lined potato beetle or the potato bug, is a major pest of potato crops. It is approximately 10 millimetres (0.39 in) long, with a bright yellow/orange body and five bold brown stripes along the length of each of its elytra. Native to America, it spread rapidly in potato crops across America and then Europe from 1859 onwards.

Taxonomy

Leptinotarsa decemlineata adult specimen.

The Colorado potato beetle was first observed in 1811 by Thomas Nuttall and was formally described in 1824 by the American entomologist Thomas Say. The beetles were collected in the Rocky Mountains where they were feeding on the buffalo bur, *Solanum rostratum*. The genus *Leptinotarsa* is assigned to the chrysolmelid beetle tribe *Doryphorini* (located in subfamily Chrysomelinae), which it shares with five other *genera*: *Doryphora*, *Calligrapha*, *Labidomera*, *Proseicela*, and *Zygogramma*.

Adult beetles average 6–11 millimetres (0.24–0.43 in) in length and 3 millimetres (0.12 in) in width. The beetles are orange-yellow in colour with ten characteristic black stripes on the elytra. The species name *decemlineata*, meaning 'ten-lined', derives from this feature. Adult beetles may, however, be visually confused with *L. juncta*, the false potato beetle, which is not an agricultural pest. *L. juncta* also has alternating black and white strips on its back, but one of the white strips in the center of each wing cover is missing and replaced by a light brown strip.

The orange-pink larvae have a large, nine segmented, abdomen and black head, prominent spiracles and may measure up to 15 millimetres (0.59 in) in length in their final instar stage. The beetle larva has four instar stages. The head remains black throughout these stages, but the pronotum changes colour from black in first- and second-instar larvae to having an orange-brown edge in its third-instar. In fourth-instar larvae, about half the pronotum is coloured light brown. This tribe is characterised within the subfamily by round to oval shaped convex bodies which are usually brightly coloured, simple claws which separate at the base, open cavities behind the procoxae, and a variable apicial segment of the maxillary palp.

Distribution

The beetle is native to North America, and is present in every state and province except Alaska, California, Hawaii, and Nevada. It now has a wide distribution across Europe and Asia, totalling over 16 million km^2.

Its first association with the potato plant (*Solanum tuberosum*) was not made until about 1859 when it began destroying potato crops in the region of Omaha, Nebraska. Its spread eastward was rapid, at an average distance of 140 km per year. By 1874 it had reached the Atlantic Coast. From 1871, the American entomologist Charles Valentine Riley warned Europeans about the potential for an accidental infestation caused by the transportation of the beetle from America. From 1875, several western European countries, including Germany, Belgium, France and Switzerland, banned imports of American potatoes to avoid infestation by *L. decemlineata*.

These controls proved ineffective as the beetle soon reached Europe. In 1877, *L. decemlineata* reached the United Kingdom and was first recorded from Liverpool docks, but it did not become established. There have been many further outbreaks: the species has been eradicated in the UK at least 163 times. The last major outbreak was in 1976. It remains as a notifiable quarantine pest in the United Kingdom and is monitored by DEFRA to prevent it from becoming established. A cost-benefit analysis from 1981 suggested that the cost of the measures used to exclude *L. decemlineata* from the UK was less than the likely costs of control if it became established.

Elsewhere in Europe, the beetle became established near USA military bases in Bordeaux during or immediately following World War I and had proceeded to spread by the beginning of World War II to Belgium, the Netherlands and Spain. The population increased dramatically during and immediately following World War II and spread eastward, and the beetle is now found over much of the continent. After World War II, in the Soviet occupation zone of Germany, almost half of all potato fields were infested by the beetle by 1950. In East Germany they were known as *Amikäfer* ('Yankee beetles') following a governmental claim that the beetles were dropped by American planes. In the EU it remains a regulated (quarantine) pest for the UK, Republic of Ireland, Balearic Islands, Cyprus, Malta and southern parts of Sweden and Finland. It is not established in any of these Member States, but occasional infestations can occur when, for example, wind blows adults from Russia to Finland.

The beetle has the potential to spread to temperate areas of East Asia, India, South America, Africa, New Zealand, and Australia.

Lifecycle

Eggs laid on the underside of a leaf.

First-instar larva after hatching.

Early (3rd) instar stage of larvae.

Colorado potato beetle females are very prolific and are capable of laying over 500 eggs in a 4- to 5-week period. The eggs are yellow to orange, and are about 1 mm (0.039 in) long. They are usually deposited in batches of about 30 on the underside of host leaves. Development of all life stages depends on temperature. After 4–15 days, the eggs hatch into reddish-brown larvae with humped backs and two rows of dark brown spots on either side. They feed on the leaves of their host plant. Larvae progress through four distinct growth stages (instars). First instars measure approximately 1.50 mm (0.059 in) long, and the last (fourth) instars measure 8 mm (0.31 in) in length. The first through third instars each last about 2–3 days duration; the fourth lasts 4–7 days. Upon reaching full size, each fourth instar spends several days as a nonfeeding prepupa, which can be recognized by its inactivity and lighter coloration. The prepupae drop to the soil and burrow to a depth of several inches, then pupate. In 5 to 10 days, the adult beetle emerges to feed and mate. This beetle can thus go from egg to adult in as little as 21 days. Depending on temperature, light conditions, and host quality, the adults may enter diapause and delay emergence until spring. They then return to their host plants to mate and feed; overwintering adults may begin mating within 24 hours of spring emergence. In some locations, three or more generations may occur each growing season.

Late (4th) instar stage of larva, before pupation.

Pupa.

Adult beetle after emergence.

Mating adult beetles.

Behavior and Ecology

Diet

L. decemlineata has a strong association with plants in the family Solanaceae, particularly those of the genus *Solanum*. It is directly associated with *Solanum cornutum* (buffalo-bur), *Solanum nigrum* (black nightshade), *Solanum melongena* (eggplant or aubergine), *Solanum dulcamara* (bittersweet nightshade), *Solanum luteum* (hairy nightshade), *Solanum tuberosum* (potato), and *Solanum elaeagnifolium* (silverleaf nightshade). They are also associated with other plants in this family, namely the species *Solanum lycopersicum* (tomato) and the genus *Capsicum* (pepper).

Predators

At least 13 insect genera, three spider families, one phalangid (Opiliones), and one mite have been recorded as either generalist or specialized predators of the varying stages of *L. decemlineata*. These include the ground beetle *Lebia grandis*, the Coccinellid beetles *Coleomegilla maculata* and

Hippodamia convergens, the shield bugs *Perillus bioculatus* and *Podisus maculiventris*, as well as various species of the lacewing genus *Chrysopa*, the wasp genus *Polistes*, and the damsel bug genus *Nabis*.

Coleomegilla maculata preying upon Colorado beetle eggs.

The predatory ground beetle *L. grandis* is a predator of both the eggs and larvae of *L. decemlineata*, and its larvae are parasitoids of the pupae. An adult *L. grandis* may consume up to 23 eggs or 3.3 larvae in a single day.

In a laboratory experiment, *Podisus maculiventris* was used as a predatory threat to female *L. decemlineata* specimens, resulting in the production of unviable trophic eggs alongside viable ones; this response to a predator ensured that additional food was available for newly hatched offspring in order to increase their survival rate. The same experiment also demonstrated the cannibalism of unhatched eggs by newly hatched *L. decemlineata* larvae as an anti-predator response.

Examples of parasitoids, predators, and pathogens of different life stages of Leptinotarsa decemlineata 1					
Type	Species	Order	Predates	Location	Reference
Parasitoid	Chrysomelobia labidomerae	Acari	Adults	USA, Mexico	
	Edovum puttleri	Hymenoptera	Eggs	USA, Mexico, Colombia	
	Anaphes flavipes	Hymenoptera	Eggs	USA	
	Myiopharus aberrans	Diptera	Eggs	USA	
	Myiopharus doryphorae	Diptera	Larvae	USA, Canada	
	Meigenia mutabilis	Diptera	Larvae	Russia	
	Megaselia rufipes	Diptera	Adults	Germany	
	Heterorhabditis bacteriophora	Nematoda	Adults	Cosmopolitan	
	Heterorhabditis heliothidis	Nematoda	Adults	Cosmopolitan	
Predator	Lebia grandis	Coleoptera	Eggs, Larvae, Adults	USA	
	Hippodamia convergens	Coleoptera	Eggs, Larvae	USA, Mexico	
	Euthyrhynchus floridanus	Hemiptera	Larvae	USA	
	Oplomus dichrous	Hemiptera	Eggs, Larvae	USA, Mexico	
	Perillus bioculatus	Hemiptera	Eggs, Larvae, Adults	USA, Mexico, Canada	
	Podisus maculiventris	Hemiptera	Larvae	USA	

	Pselliopus cinctus	Hemiptera	Larvae	USA	
	Sinea diadema	Hemiptera	Larvae	USA	
	Stiretrus anchorago	Hemiptera	Larvae	USA, Mexico	
Pathogen	Bacillus thuringiensis subsp. tenebrionis	Bacteria	Larvae	USA, Canada, Europe	
	Photorhabdus luminescens	Bacteria	Adults, Larvae	Cosmopolitan	
	Spiroplasma	Bacteria	Adults, Larvae	North America, Europe	
	Beauveria bassiana	Hypocreales	Adults, Larvae	USA	

Agricultural Pest

Dutch newsreel.

Potato Crop Pest

In about 1840, *L. decemlineata* adopted the cultivated potato into its host range and it rapidly became a most destructive pest of potato crops. They are today considered to be the most important insect defoliator of potatoes. They may also cause considerable damage to tomato and eggplant crops with both adults and larvae feeding on the plant's foliage. Larvae may defoliate potato plants resulting in yield losses of up to 100% if the damage occurs prior to tuber formation. Larvae may consume 40 cm^2 of potato leaves during the entire larval stage, but adults are capable of consuming 10 cm^2 of foliage per day.

The economic cost of insecticide resistance is significant, but published data on the subject is minimal. In 1994, total costs of the insecticide and crop losses in the US state of Michigan were $13.3 million, representing 13.7% of the total value of the crop. The estimate of the cost implication of insecticides and crop losses per hectare is $138–368. Long-term increased cost to the Michigan potato industry caused by insecticide resistance in Colorado potato beetle was estimated at $0.9 to $1.4 million each year.

Insecticidal Management

The large scale use of insecticides in agricultural crops effectively controlled the pest until it became resistant to DDT in 1952 and dieldrin in 1958. Insecticides remain the main method of pest control on commercial farms. However, many chemicals are often unsuccessful when used against this pest because of the beetle's ability to rapidly develop insecticide resistance. Different populations in different geographic regions have, between them, developed resistance to all major classes of insecticide, although not every population is resistant to every chemical. The species as a whole has evolved resistance to 56 different chemical insecticides. The mechanisms used include improved metabolism of the chemicals, reduced sensitivity of target sites, less penetration and greater excretion of the pesticides, as well as some changes in the behavior of the beetles.

Examples of insecticides available for the control of Colorado potato beetle on different crops in Kentucky, USA.					
Insecticide class	Common examples	Potato	Eggplant	Tomato	Notes
Organophosphates	phosmet	X			on US Emergency Planning List of Extremely Hazardous Substances
	disulfoton	X		X	Usage restricted by US government; manufacturer Bayer exited US market 2009
Carbamates	carbaryl	X	X	X	Widely used in US
	carbofuran	X			One of the most toxic carbamates
Chlorinated hydrocarbons	methoxychlor	X		X	Banned in EU 2002, in USA 2003
(Cycloldienes)	endosulfan	X	X	X	Acutely toxic, bioaccumulates, endocrine disruptor. Global ban 2012 with exemptions until 2017
Insect growth regulator	azadirachtin	X	X	X	
Spinosin	spinosad		X	X	
Avermectin	abamectin	X		X	

Non-pesticidal Management

East German Young Pioneers collecting beetles during the war against the potato beetle.

Bacterial insecticides can be effective if application is targeted towards the vulnerable early-instar larvae. Two strains of the bacterium *Bacillus thuringiensis* produce toxins which kill the larvae. Other forms of pest control, through non-pesticidal management are available. Feeding can be inhibited by applying antifeedants, such as fungicides or products derived from Neem (*Azadirachta indica*), but these may have negative effects on the plants as well. The steam distillate of fresh leaves and flowers of tansy (*Tanacetum vulgare*) contains high levels of camphor and umbellulone and these chemicals are strongly repellent to *L. decemlineata*.

Beauveria bassiana (Hyphomycetes) is a pathogenic fungus that infects a wide range of insect species, including the Colorado potato beetle. It has shown to be particularly effective as a biological pesticide for *L. decemlineata* when used in combination with the bacterium *Bacillus thuringiensis.*

Crop rotation is, however, the most important cultural control of *L. decemlineata*. Rotation may delay the infestation of potatoes and can reduce the build-up of early season beetle populations because the adults emerging from diapause can only disperse to new food sources by walking. One 1984 study showed that rotating potatoes with non-host plants reduced the density of early season adults by 95.8%.

Other cultural controls may be used in combination with crop rotation: Mulching the potato crop with straw early in the growing season may reduce the beetle's ability to locate potato fields, and the mulch creates an environment that favours beetle's predators; Plastic-lined trenches have been used as pitfall traps to catch the beetles as they move toward a field of potatoes in the spring, exploiting their inability to fly immediately after emergence; Flamethrowers may also be used to kill the beetles when they are visible at the top of the plant's foliage.

Relationship with Humans

Cold War Villain

During the Cold War it was claimed by some countries in the Warsaw Pact that the beetles had been introduced by the CIA in an attempt to reduce food security by destroying the agriculture of the Soviet Union. A widespread campaign was launched against the beetles; posters were put up and school children were mobilized to gather the pests and drown them in benzene or spirit.

Philately

Statue of the Colorado potato beetle in Hédervár, Hungary. It marks the discovery of the beetle at the site in 1947 during the rapid spread of the pest in Europe throughout the 20th century.

L. decemlineata is an iconic species and has been used as an image on stamps because of its association with the recent history of both North America and Europe. For example, in 1956, Romania issued a set of four stamps calling attention to the campaign against insect pests and it was featured on a 1967 stamp issued in Austria. The beetle also appeared on stamps issued in Benin, Tanzania, the United Arab Emirates, and Mozambique.

In Popular Culture

During the 2014 pro-Russian unrest in Ukraine, the word *kolorady*, from the Ukrainian and Russian term for Colorado beetle, gained popularity among Ukrainians as a derogatory term to describe pro-Russian separatists in the Donetsk and Luhansk Oblasts (provinces) of Eastern Ukraine. The nickname reflects the similarity of black and orange stripes on so-called St. George's ribbons worn by many of the separatists.

Potato Bug

A potato bug, sometimes referred to as a Jerusalem Cricket or Child of the Earth, is a small several inch long insect found in organic material such as potatoes and other crops. The potato bug likes to live in moist conditions and has an unusually large head compared to its body and emits an unpleasant odor.

Actually, the term potato bug can refer to two different insects. The strange, alien-looking potato bug is also known as a Jerusalem cricket. The Jerusalem cricket, scientific name Stenopelmatus fuscus, is a large wingless insect found throughout portions of the southwest United States. Meanwhile, there's another potato bug, the Colorado potato bug, scientific name Leptinotarsa decemlineata.

The Colorado potato bug is also found throughout the Southwest US, but it has since distributed itself far and wide, and can now be found all across the US and throughout much of Europe and Asia. Both kinds of potato bugs have different characteristics, but they both feed on potatoes, naturally.

The Jerusalem Cricket

- Length: 5.08 to 6.35 Centimeters (2 to 2.5 Inches).
- Native to US Southwest.
- Diet: Insects, fruits, plant roots.

The Jerusalem cricket really does have a bizarre, alien-esque appearance to it. They have long, yellowish thoraxes with abdomens that typically have stripes or splotches which alternate between black and brown. Jerusalem crickets can grow up to two and a half inches long, and they have long spear-like legs that they dig into soil with. Complicating the already hazy distinctions of the term potato bug, the term Jerusalem cricket is often used to refer to a wide variety of different insects in the Stenopelmatidae family.

Despite their name, Jerusalem crickets aren't native to Jerusalem, and they are in fact mainly found in the western half of the United States. (Incidentally, they aren't true crickets either.) They can be found throughout Arizona, New Mexico, and all along the Pacific coast of the country, being found all the way from British Columbia down to Baja California and many other parts of Mexico. Habitat wise the Jerusalem cricket mainly lives in underground burrows, but they can be found above ground in warm damp places.

The Jerusalem cricket is pretty harmless to humans, as much as its large size tends to unnerve people. The insect isn't poisonous, although it can leave a painful welt with its bite. While the Jerusalem cricket isn't a true cricket, like a cricket it does produce a song when mating. The Jerusalem cricket bumps its body against the ground, producing a drumming rhythm that is intended to attract mates. The insects can also hiss to ward off potential predators, though they don't do it like other hissing insects do. Other hissing insects typically hiss by forcing air through their abdomen, just like hissing cockroaches do. By contrast, the Jerusalem cricket produces a scratchy hissing noise by scraping its hind legs across its abdomen.

Though Jerusalem crickets have been dubbed potato bugs because of their tendency to eat potatoes, potatoes are actually not their preferred diet. Jerusalem crickets actually prefer to eat smaller insects, fruits, and the roots of plants other than potatoes. That said, the Jerusalem crickets can indeed be found eating the tubers and roots of potato, though far less frequently than their name would imply. They have occasionally done damage to potato fields, but they aren't considered a primary pest to potato plantations.

Jerusalem crickets are capable of causing severe damage to crops if large groups of them descend upon a potato field though. Pesticides are one of the primary ways farmers deal with the bugs, though trapping and baiting them is also effective. The Jerusalem crickets can also be deterred by removing objects which they can hide in from a field, such as rock piles or wood piles.

Colorado Potato Bug

- Length: 1.27 Centimeters (0.5 Inches).
- Native to most of the US, can now be found in Asia and Europe.
- Diet: Fruit and vegetable leaves, eggplants, tomatoes, potatoes.

By contrast to the odd look of the Jerusalem cricket, the Colorado potato bug looks much more like a typical beetle. They are around half an inch long and have black vertical stripes on their yellow bodies. They're oval shaped, and unlike the Jerusalem cricket, they are capable of flying.

The native range of Colorado potato beetles covers most places in the continental United States except for California and Nevada. As mentioned, the beetles have spread far beyond this original territory and can now be found in most of Asia and Europe. The Colorado potato bugs stay deep underground for the cold winter months, emerging during the spring to lay eggs on the underside of leaves.

The Colorado potato bugs become active during the spring at about the same time potatoes begin to emerge from the ground and grow their tubers. The adult insects stick around the potato fields to feed for a little while, and then begin to mate with one another. A female Colorado potato bug is capable of leaving behind 10-30 eggs on a single leaf. The lifetime of an adult female Colorado potato bug is typically a few weeks long, depending on variables like temperature and humidity, and during this time the females can lay up to 350 eggs each.

While the Jerusalem cricket isn't considered a major pest to potato farms, the Colorado potato bug really does damage potato crops quite frequently. The bugs feed on the potato plants and their lifecycle is dependent upon them. During the spring the potato bugs lay eggs on the underside of potato leaves. When the larvae hatch from the eggs, they then feed upon the leaves of the potato plant. They can also feed on other plants like eggplants, peppers, and tomatoes.

Colorado potato bugs can be a very difficult pest to manage. Widespread and indiscriminate use of pesticides has created populations of potato bugs which are resistant to pesticides. For this reason, potato farmers should use integrated pest management plans. The bugs can be handpicked off of plants if they are in a smaller garden. Using natural predators of the bug like birds of the shrike family and ladybugs can help keep their numbers in check. Rotating crops and using microbial defenses can also help reduce overall numbers of the pest. Researchers are even exploring genetic engineering as a possible defense for crops.

Livestock Pests

Thousands of pests affect our major livestock species, with impacts on livestock production varying in space and time. Livestock pests and parasites feed on the blood, skin, hair, feathers, and wounds of an animal. Their bites cause physical and mental health issues for the animal. They are also vectors of disease.

Cattle Pests

Horn Fly

The horn fly is a common pest of range and pasture cattle throughout the Pacific Northwest. Economic numbers of horn flies develop in June, and the flies are abundant during July, August, and September. Many cattle ranchers have found chemical control of horn flies essential for good herd management.

The small, bloodsucking flies cluster on the back, sides, and bellyline of all cattle breeds. A thousand or more flies may infest an individual animal. Feeding by the flies on animals results in considerable discomfort and annoyance, in addition to the loss of blood. Tormented cattle toss their heads, switch their tails, and rub through brush in an attempt to lessen the attacks of these flies. Heavily infested cattle do not graze during the day and "brush up" for long periods to avoid fly attack.

Single grazing season studies with yearling heifers and steers have shown increased weight gains of 15 to 50 pounds per animal when horn flies are controlled. When cows are heavily infested with horn flies, milk production may be reduced 10% to 20%.

Identification. The horn fly is about half the size of the common house fly. It is light gray and about 3/16 inch in length. It has piercingsucking mouthparts. At rest, the head of the horn fly usually points downward. The wings are held flat over the back at an angle of about 60 degrees. The fly burrows among the hairs on the host to feed.

The flies are usually observed on the backs of cattle when the weather is cloudy or cool.

Biting Flies of Cattle

During sunny weather the air temperature seems to determine their location on an animal. As the air temperature rises in the morning, the flies are first observed on the sunny side of the animal, then on the shaded side and later on the under-side along the animal's bellyline.

As the air temperature lowers during the late afternoon and evening, the flies return to the sides and back of the animal. When disturbed, they will fly up in a swarm but return quickly to individual animals.

Life Cycle. Adult flies spend their entire life on the host, except when the females leave to deposit eggs on fresh, undisturbed cattle droppings. The eggs hatch in approximately 24 hours. The maggots complete their development in the droppings within 4 to 5 days and the mature maggot transforms into an adult fly in 5 to 7 days. The life cycle from egg to adult usually takes 10 days to 2 weeks depending on temperature and moisture conditions. Many generations overlap each summer. Horn flies spend the winter as maggots or pupae beneath cattle droppings.

Stable Fly

Stable flies are distributed throughout the Pacific Northwest. Mostly a pest around fenced pastures and feedlots, they generally are not a problem in range situations. They are most abundant from midsummer through autumn.

These flies are vicious biters. They frequently change positions or move to several animals to feed. The behavior of interrupted feeding increases their annoyance factor. Stable flies feed mainly on the legs of cattle and appear resting on sunny or light colored surfaces.

Identification. Similar in appearance to house flies, stable flies are brownish gray with broken stripes on the thorax. The abdomen is checkered in appearance and shorter than the abdomen of the house fly. Wings of stable flies when at rest are held in an angular position similar to horn flies. Their prominent proboscis protrudes forward in front of the head.

Life Cycle. The female lays small white eggs in decaying organic matter. Eggs sometimes will be laid in manure mixed with straw but soggy hay, grain, or straw around feed bunks are preferred for oviposition. In 3 days the eggs hatch into typical maggot-type larvae. After feeding for 2 to 3 weeks the larvae crawl to drier areas of the food source and pupate. Adult flies emerge from the pupa in 9 to 13 days. The egg to adult period averages 35 days. Most adults live around 3 weeks during the summer. Winter is passed as a larva or pupa.

Tabanids, Mosquitoes and Blackflies

Several different kinds of tabanids (horse flies and deer flies), mosquitoes, and blackflies (buffalo gnats) attack cattle in summer pastures or under range conditions. A few horse flies per animal can excite cattle and cause considerable trouble, whereas cattle generally tolerate fairly large numbers of mosquitoes and blackflies. Actual weight loss, blood loss, or other damage caused by these flies has not been measured in Washington. However, various studies in other areas show that losses can be considerable when livestock are under heavy attack. Control measures specifically for these insects have been impractical to date. However, controls for horn fly will give some control of tabanids, blackflies, and mosquitoes.

- Tabanids: Horse fly and deer fly larvae develop in damp to wet soil at the edges of streams and lakes and in marshy areas. Larvae range from saprophytic to predaceous. Predaceous species feed on other insects, earthworms, snails, and crustaceans. Usually 1 year is spent in the larval stage with adults emerging from late spring throughout the summer, depending on the species. Like mosquitoes, only female tabanids feed on blood and bite only during the daytime. Adult females live about 2 weeks to a month, depending on temperature. They are strong fliers and may be bothersome a mile or more from their breeding sites.
- Mosquitoes: Most adult female mosquitoes must have blood meals to develop eggs. The eggs are laid in or near water and the larvae live in various aquatic habitats. Some species of mosquitoes develop in snow melt pools in mountain valleys and meadows. Some develop in overflow or seep pools next to creeks, rivers, or irrigation canals. Others are found in septic tank overflows, rain puddles, and water-filled artificial containers. Eggs are laid on the water surface or moist soil. Once in the water or covered with water (in the case of eggs laid on the soil) the eggs hatch in 3 to 4 days. Larval development takes from 5 days to 3 weeks. The pupal stage lasts 2 to 4 days before adults emerge. Mosquito adults live 2 to 4 weeks or more, depending on temperature. Mosquitoes bite during day or night, depending on species.
- Blackflies: Buffalo gnats breed in running water and are present in streams as well as certain irrigation canals. Eggs are laid on rocks or trailing vegetation or dropped onto the water surface. Eggs hatch within a few days. Depending on the species, water temperatures, and food availability, larval development takes from a month to a year. Most species in mountain streams have one or two generations per year. Those species breeding in the warmer irrigation canals can produce three to five generations per year.

Large numbers of blackfly adults have killed livestock in Canada due to blood loss and toxemia. Adult blackflies bite only in the daytime. Under moderate blackfly attack, cattle will lie down, "brush up," or stand in water to avoid blackfly attacks. Feeding is interrupted and milk production decreases. Blackflies are particularly hard to control because the adults can move far from breeding sites. Control generally uses chemicals to kill the larvae. This type of program has been successful in certain states for controlling blackflies attacking sheep. Cattle in the treated areas also are protected.

Other Biting Flies

Other biting flies such as punkies and snipe flies bite cattle and other livestock. Little is known about the economic importance in general except that if they were numerous they probably would cause many of the weight loss problems, etc., that our common horn flies, etc., cause. Some of the punkies do act as vectors for the infamous blue tongue virus.

Cattle Grubs

Cattle grubs, Hypoderma sp., cost the beef industry over 200 million dollars annually. Besides cost of control, losses result from poor gain by infested animals or from paralysis which may occur when grubs migrate through the spinal canal. Occasionally secondary infection may develop in grub emergence holes. On slaughter animals, grub damage in loin muscles causes trim loss and downgrading of carcasses. Hides of grubby cattle are peppered with holes in the thickest, most valuable part of the hide. The adult stage of cattle grubs causes economic losses from gadding the cattle (running from the insects) because the animals do not graze properly or can injure themselves running, and from milk reduction for calves.

Life Cycle

The adults are known as heel flies or warble flies. In the Pacific Northwest, they generally appear in early summer. Two species are involved. The northern cattle grub lays its eggs singly on hairs on the lower part of the animal's body. The common cattle grub attaches her eggs in the same areas, but may attach as many as a dozen eggs to a single hair. Adult flies do not feed so they live only 3 to 10 days. Cattle will often flee in panic from an egg laying female cattle grub. This is called "gadding." We do not know why cattle fear the fly because it can't bite or sting.

The eggs hatch in about 4 days, and the tiny larvae crawl down the hair and penetrate the skin. Where several larvae have entered the skin at the same place, a reaction called "hypodermic rash" develops. Infected animals constantly lick and scratch these spots and do not feed normally. The grubs suffer very high natural mortality during their development because of the natural resistance of older hosts. Cattle of all breeds are the normal host, but occasionally horses are attacked. Calves and yearlings tend to have more grubs than older stock.

Little is known about the larvae after it enters the animal's body but in 1 to 2 months the small common cattle grubs have migrated to just beneath the lining of the esophagus. Small northern cattle grubs appear in the spinal canal after that same time interval.

The grubs live in those locations for about 6 months (during the winter) and then migrate to the topline of the animal. When the larva first arrives in the back, it cuts a small breathing hole in the hide, then molts in a pouch a warble between the inner and outer layers of skin. After growing rapidly for 6 to 8 weeks, the fully grown larva enlarges the hole, exits the warble and falls to the ground. There it crawls under some object and transforms to a pupa. The pupal stage lasts 2 weeks to 60 days depending upon the weather. The adult female then emerges, mates, finds a host, and lays eggs. Male cattle grubs occur at topographic aggregation sites.

Face Fly on Beef Cattle

The face fly is a pest of range and pasture livestock in most areas of Washington. This fly poses a serious problem to cattle because it is associated with a high incidence of eye disorders.

This nonbiting fly feeds on the moisture and secretions about the eyes and muzzle of cattle, horses and sheep. Its feeding habits cause increased tear production and eye swelling. The face fly is involved in the mechanical transmission of disease organisms to the eye. In addition, it will feed upon blood or exudate (ooze) from fresh wounds and saliva deposits.

The presence and feeding of this persistent insect also cause extreme annoyance and irritation to its hosts. Infested livestock may bunch together or seek dense shade and refuse to feed. Face flies usually are not a pest of confined cattle unless they are adjacent to pastured cattle. The face fly is also a household pest because of its habit of hibernating in attics and interwall areas of houses, barns, churches and other buildings. On warm days during the winter and early spring, they move into living areas, sometimes in great abundance, and become active pests.

Habitat and Life Cycle

Face flies were first found in Washington in the mid-1960s. They now occur throughout the state. Natural enemies have stabilized the populations since that time. However, because face fly populations can increase rapidly, check their numbers on cattle often.

Face flies are troublesome on livestock during most of the summer. The flies prefer bright sunshine and usually do not follow livestock into dense shade or buildings.

Eggs are laid only in undisturbed, fresh cattle droppings. Face fly maggots develop in the droppings, pupate in dry areas of droppings or in the surrounding soil, and emerge as adults. Development from egg to adult requires 10 to 21 days under ideal conditions. Successive generations occur during the warm months of the year.

Cattle Lice

Lice cause losses in all types of beef cattle operations. Animals that develop extremely large louse populations may abort or die from anemia caused by blood loss, but the major loss results from poor feed conversion. Lice are primarily a wintertime problem because self-grooming, sun-shine, rain, and short summer hair coat make poor conditions for their development. Cattle breeds also differ in susceptibility.

In cow-calf operations, lousy cattle will produce less milk and will wean lighter calves. Cattle ranchers will find that lousy cattle require more feed for each pound gained. Louse-infested cattle appear unthrifty and have dry looking skin. They also damage fences, corrals, and feeder bunks by rubbing on them to relieve the irritation caused by lice feeding. Cattle are the only host of cattle lice. Usually, 1% or 2% of the cattle in a herd are carriers that will harbor lice year-around. Bulls are often carriers since they have a denser coat and cannot self-groom as easily as the other cattle. Lice spread by contact from carriers to other animals in the herd. Cattle lice spend their entire life on the animal and will live only a few days off the host if removed. Lice can pass from egg to adult in 3 to 4 weeks. Eggs that fall off the host will not hatch unless the weather is hot and then the young must find a bovine host within 2 days or they will die.

Each time cattle are handled in winter and early spring, they can be quickly examined for lice. Part the hair with your fingertips and examine the animal in several places, including the neck, withers, brisket, shoulders, midback, tailhead, and behind the rounds.

The Pacific Northwest has four species of lice that infest cattle. Three of these species feed by piercing the skin and sucking blood. The shortnosed cattle louse is normally found as a pest of adult animals. The longnosed cattle louse infests calves most heavily but often is found on mature animals in low numbers. The little blue cattle louse is found on all ages of cattle. The fourth species, the cattle biting louse, feeds on the scales of the skin and causes irritation. It is reddish instead of blue like the bloodsucking lice, and it usually is found at the base of the tail or withers of cattle of all ages.

Scabies on Cattle

Scabies, or scab mange, is a skin condition caused by the feeding of three kinds of tiny, parasitic mites in the skin of cattle and other animals. The disease takes its name from the scabby appearance of infested animals. Body fluids ooze from the feeding wounds caused by the mites and an obvious dermatitis develops. Mites are active on cattle all year, but the scabby condition they cause is most severe in winter.

Another reaction of the cattle to mite infestation includes a generally unthrifty condition. Up to 100 additional days feeding may be required to bring infested cattle to marketable condition. Cattle stressed by a mite infestation also tend to be more prone to respiratory problems, such as pneumonia. When infested cattle are slaughtered, scars on the hides result in lower hide prices.

Scabies is quarantinable and cattle exposed to infested livestock or facilities also may be quarantined. Quarantine regulations require cattle workers not only to dip infested cattle, but also to disinfect and clean corral facilities and trucks used to haul infested animals. Complying with overall quarantine regulations is a burden that is very costly.

Quarantine and Eradication

Cooperative state and federal programs have reduced the incidence of scabies through quarantine and eradication. All cases of scabies reported to state regulatory agencies are investigated to determine the source of the infestation and the possibility of contact of infested animals with noninfested animals. When an infestation involves movement of animals across state lines, the Animal and Plant Health Inspection Service of USDA determines the proper course of action for management of a potential outbreak. Veterinarians from state and federal regulatory agencies can provide specific information on quarantine and eradication programs.

Types of Scabies

Three types of cattle scabies occur in the West, and each is caused by a different mite— the psoroptes, sarcoptes, and the chorioptes. Cattle may be infested with any one species or combinations of these tiny mites. These mites are barely large enough to be visible to the unaided eye against a black background. Each requires 10 to 12 days to complete a life cycle and each lives on or within the skin tissues of the animal where feeding and reproduction occur.

The most severe form of scabies is caused by psoroptic mites. The initial infection site for psoroptic scabies is usually where the hair is thick, as along the top-line of an animal from the withers to tailhead. Sarcoptic mites frequently will first infest an animal above the rear portion of the udder or scrotum or on the inner surfaces of the thighs where the hair is very thin. Chorioptic mites, the cause of the most common form of scabies, also have a tendency to produce their first lesions in these same areas.

Poultry Pests

Lice, fleas, and mites in particular are the most common pests in poultry in Washington. Infested birds have scaly, scabby skin and discolored or ragged feathers. They also lose many feathers. They also may have deformed feet.

Consult current local pesticide recommendations when a pest problem occurs. Well planned and constructed poultry houses and good sanitation make control easier.

Poultry Mites

Chicken Mite. During the daytime these mites stay in cracks, crevices, and other protected places. At night they crawl upon the birds to feed. Their only food is the blood of the birds they infest.

Since these mites rarely stay on the birds during the daytime, a flock may be badly run down without the owner being aware of the cause of the trouble. In cases of heavy infestations, the birds become droopy, pale about the head, and listless. They may stop laying.

Northern Fowl (Feather) Mite. This mite resembles the chicken mite, but it breeds among the feathers and spends its entire life on the bird. The presence of the mites often is first noticed on the eggs. Mite eggs and excrement will be found among the feathers, especially about the vent and tail.

These mites cause annoyance, loss of blood, and unthriftiness in infested birds. English sparrows also are infested with these mites. Destroying their nests and spraying their nest sites around poultry houses is suggested.

Depluming Mite. This mite lives at the base of the feathers. It appears to produce an intense itching, causing the birds to pull out their feathers. If you examine the stumps of the feathers soon after the quill is broken, you will find them surrounded by scales and crusts. These mites become active in the spring and usually disappear in winter.

Scaly Leg Mite. This form of itch mite causes a condition known as scaly leg. While it usually occurs only on the legs, the mite may attack the combs and wattles. The infestation is easy to recognize by the enlargement of the feet and legs and their rough appearance. In severe cases, the joints become inflamed and the birds become lame.

Poultry Lice

Lice on poultry do not suck blood, but feed by nibbling or chewing dry skin, scales, feathers, or scabs on the skin. Several kinds of lice attack birds in Washington. Some of the most important are the chicken body louse, the shaft louse, and the chicken head louse.

- Chicken Body Louse: This louse lives most of the time on the skin, mainly in the areas around the vent and under the wings. It is common on both young and old birds.
- Shaft Louse: The shaft louse, or common body louse, is similar to the chicken body louse, but smaller and paler colored. It is common around the vent as well as on the back and breast. It tends to cling to the feathers rather than to the skin of the bird. This louse does not infest young chickens.

- Chicken Head Louse: This louse is a problem mostly on young birds. It occurs on the head at the base of the feather or down. It passes from one chicken to another by contact.

Fleas

Several kinds of fleas may occasionally infest birds, but only one kind, the western hen flea, is a problem in Washington. It breeds in the soil, principally in droppings, and does not infest birds except for brief feeding periods.

Darkling Ground Beetle

The "mealworm" stage of this beetle occasionally burrows into the flesh and internal organs of dead or dying chicks of 2 or 3 weeks of age. This insect has also been shown capable of transmitting leucosis in broilers.

Full-grown larvae are brown banded, sausage shaped, more pointed at one end, and about 3/16 inch long. The adult beetles are the same length and brown-black in color.

Livestock Ticks

Ticks are small animals, similar to insects and spiders. In an unfed condition, their bodies are oval, flattened, and appear hard and leathery. Adult ticks have eight legs; the larvae or "seed" ticks have only six. Larval ticks are very small, about 1/40 inch long. The intermediate or nymphal stage is about 1/16 inch; adults are about 3/16 inch. Fully engorged females of common species can expand to 1/2 inch in length and are more spheroid.

Ticks feed exclusively on the blood of animals, piercing the host's skin with their mouth parts. A series of barbs along the shaft of the mouth parts anchor the head of the tick to the host's skin, making removal difficult. Unless the tick is removed, it remains feeding from several hours to many days before becoming fully engorged. Only then will it release its hold and drop from the host to molt or lay eggs.

Two major kinds of ticks live in Washington—host ticks (Ixodidae) and nest ticks (Argasidae).

Three common kinds of host ticks in Washington bite humans and certain livestock. They are Dermacentor andersoni (the Rocky Mountain wood tick), Dermacentor variabilis (the American dog tick), and Idoxodes pacificus (a tick found in western Washington). These species are usually encountered outdoors in brushy, woodland areas. They feed on a variety of mammals, large and small. The brown dog tick, Rhipicephalus sanguineus, is primarily a pest of dogs and only rarely bites humans. It lives in dog kennels, houses, and other areas where dogs are kept, including human habitations.

Life Cycle

The life cycles of all species of ticks are similar, although the host range and duration of life stages vary. All are multi-host ticks. Three separate blood meals, generally from hosts of a different size, are required to complete the stages of development. The tick leaves the host at the end of each blood meal to molt or lay eggs. The brown dog tick also requires three blood meals, but the dog serves as host for each meal.

Once an adult female has mated and become fully engorged with blood, she drops from the host to seek a protected place to lay eggs. Thousands of eggs are laid in a period of 10 days to 3 weeks. The eggs hatch in about a month and the small six-legged larvae crawl up on vegetation to await a passing host.

The larvae host ticks usually feed on smallrodents. They feed for a few days, drop fromthe host, and seek shelter for the molt tothe nymphal stage. After transforming intoan eight -legged nymph, thc tick crawls ontovegetation and attaches to a passing animal.This stage can be found on larger hosts, suchas jack rabbits. The nymph feeds for a fewdays, drops from the host, and transform-sinto an adult. Adult ticks are common on deer,horses, cattle, sheep, and other large animals.The adult stage most often attacks people.

The life cycle can last up to 3 years, although 2 years is more common. Unfed nymphs and adults persist throughout the winter and seek new hosts in the spring. A blood meal is necessary before the tick can transform into the next growth stage, or before the adult female can lay eggs. The common species of ticks can live several years without feeding.

Nest ticks are more secretive, and most species feed on birds and rodents. They also will feed on humans. Nest ticks attach and feed very rapidly so they are rarely detected. They feed on resting humans and livestock and will crawl considerable distances to reach a nesting animal. Molting takes place after a complete blood meal. Because of the short time of contact with the host, these ticks are concentrated in bedding or nest sites of mammals and birds.

Seasonal Abundance

Ticks are most active in spring when vegetation begins to grow. By July or August (in higher elevations), adults or nymphs, which have not found a host, seek a protected site to avoid summer heat. Ticks are most abundant in brushy areas where vegetation is sufficient to support both large and small animals. They concentrate near trails and openings where they are likely to encounter passing hosts.

The brown dog tick differs in its pattern of seasonal activity because it is closely associated only with canine hosts. All of the life stages can be found at any time of the year in buildings used by dogs. When not feeding, the ticks hide in cracks, corners, behind picture frames or moldings of buildings where dogs are housed. A favorite site for feeding on dogs is between the toes.

Disease Associated with Ticks

Tick bites are potentially dangerous to humans both from the effects of the bite itself and from transmitted microbes that cause disease. Throughout the United States, Rocky Mountain Spotted Fever (RMSF) is transmitted by the Rocky Mountain wood tick, the American dog tick, and to a lesser extent by the brown dog tick. In Washington, only two or three human cases of RMSF occur annually. The brown dog tick and the American dog tick also can spread malignant jaundice to canines. Tularemia or "rabbit fever" is transmitted to humans by the American dog tick and the Rocky Mountain wood tick.

Nest ticks also transmit disease agents. Humans usually are exposed when sleeping in cabins or outdoor shelters that also are used by wild rodents. These ticks transmit Relapsing Fever. Nest ticks also transmit a number of agents of poultry diseases. The bite of nest ticks also can be dangerous. Many people show extreme local hypersensitivity to the bite of these ticks.

The bite of the tick itself can cause dermatitis (an inflammation of the skin), secondary infections, and in the case of host ticks, a condition known as tick paralysis. Instances of tick paralysis are rare. It is caused by the feeding of a female tick. The tick introduces a toxin (Ixovotoxin) while feeding, which causes an ascending type of paralysis. Early removal of the tick results in the rapid and complete recovery of the affected animal or person. Failure to remove the tick results in progression of the ascending paralysis until death.

Horse Pests

Horse Bots

Horse bots are the larvae of flies that have become highly specialized as parasites of horses. Three species of horse bots attack horses in the Pacific Northwest—the common horse bot, the throat bot, and the nose bot.

Bot flies have four stages, the adult, egg, larva or bot, and pupa. In the western states, the adult flies appear about June and remain active until frost. During this time, the shortlived adult flies follow their sole purpose— reproduction. The female fly hovers and darts about the horses and lays her eggs, gluing them to the hairs of the horse's body. Each species has its preferred laying site. It is at this time that many horses react to the bot flies in a characteristic manner. The reaction varies somewhat among horse breeds.

Nose or throat bots cause horses to walk about throwing their heads violently; if this fails, the horses begin to run seeking shade or shelter. If the horses are being ridden when flies are active, they become skittish, unmanageable, sometimes running away or throwing the rider.

Most eggs of the common bot are glued onto the hairs of the lower front legs of the animals. The eggs hatch several days after they are laid, but hatching does not take place until they are rubbed by the warm moist lips of the horse. The eggs may remain for several weeks, but as soon as this stimulus occurs the larvae emerge and attach themselves to the mucous membranes of the mouth. The larvae then penetrate the mucous membranes of the tongue and gums. They burrow in the gums for about 1 month and finally migrate to the stomach where they attach to the lining of the stomach by means of strongly developed mouth hooks.

Eggs laid by the throat bot fly are deposited on the hairs of the throat and lips. The friction stimulus is not required to cause the eggs of the throat bot to hatch. The larvae hatch and crawl into the animal's mouth, lodging between the teeth and in gum pockets in the mouth. In about 1 month, the larvae move into the stomach and attach to the lining of the stomach.

The female of the nose bot fly darts in and lays her eggs on the hairs of the lips and mouth. The eggs hatch in a few days and burrow in the mucous membrane of the lips and gums; here they stay for about 1 month. The larvae then migrate to the stomach and attach to the lining of the stomach. The majority of the larvae attach to the lining near the pylorus, or exit of the stomach into the intestine.

Larvae of all three species of horse bots require about 10 months to develop, eventually passing out of the intestine in the feces. Pupation follows either under or in the fecal pile.

The larvae of the nose bot will attach themselves in the rectum and anal region before finally dropping to the soil to pupate. At this time the horse will exhibit discomfort by rubbing or switching its tail.

Much of the damage by bots is not recognized, yet the damage caused in the gums and to the digestive tract is important to the health of horses. Treatment is generally administered no earlier than 1 month after the first killing frost. Greatest success is obtained by treating in early winter, either in December or January. Horses of higher value may be treated more frequently. Eggs of the common horse bot can be washed weekly with warm water and a sponge to cause hatching and thus prevent infection.

Blood Sucking and Nuisance Flies

Certain flies including mosquitoes will annoy horses on the pasture or in the barn lot or the arena. They cause pain and discomfort to the horse, at times, making them unmanageable while they are being ridden or shown.

Stable Flies

Stable flies breed in decaying vegetable matter around the farmstead. Adult flies feed on the blood of mammals. They normally roost on walls, trees, or fences in the shade. Once or twice a day they come to an animal, feed vigorously for a minute and then leave to roost on the building and digest their meal. Feeding by stable flies causes animals much discomfort and causes them to switch and kick viciously. Cleanup and sanitation serve to break up breeding areas and contribute a good deal toward fly control. For stable flies, one can use a good residual spray on the buildings and fences to kill resting flies as they roost.

Horse Flies and Deer Flies

Horse flies and deer flies are large robust flies that suck blood from a wide range of mammals. Their larvae appear in moist soil or in standing or running water that is fairly shallow. The bite of the adult is extremely painful and causes horses a great deal of discomfort; however, they are not as numerous in the Pacific Northwest as the other flies.

Face Flies and Nuisance Flies

Face flies do not suck blood, but are a severe nuisance to livestock. They sit on the mucous membranes of the eyes and nostrils to feed. This feeding sets up a local irritation and considerable discomfort to horses and domestic livestock. Other nuisance flies or blood sucking flies are a discomfort to horses, either on pasture or while being ridden. Some will feed on blood oozing from bites of horse flies. All of these flies, except mosquitoes and punkies, attack horses outdoors only.

A number of prepared repellent mixtures may be applied to horses to ward off the attack of nuisance flies. These preparations are applied as sprays, "wipe-ons," or aerosol "bombs."

A single application just before riding is usually enough. One should be careful not to over-apply these materials. This is not from the standpoint of the chemical pesticide, but the oil with which it is formulated. Overuse of oil-based pesticides could cause skin irritation, and the animal may develop a dandruff condition. In more advanced cases the skin may become blistered, and some hair loss results. While the repellent preparations are effective, use no more than the label recommends.

Mosquitoes

Mosquitoes are a special discomfort to horses on pastures, especially during early spring in wet years. Mosquitoes may be severe enough to warrant keeping horses in the barn at night and turning them out only during daylight hours.

Using aerial spraying on the pasture and surrounding area with a registered pesticide will alleviate mosquitoes and nuisance flies for a short time per spraying. Remove horses from the pasture at the time of spraying because of their fear of the applicator plane. Because most mosquitoes do not move far from aquatic breeding sites, an effective control measure is breeding site reduction or elimination.

With the onset of West Nile Virus across the country and the implications for equine health, it is necessary to briefly address the disease here. While certain birds are the primary host, there are mammals that are dead-end hosts, some of which are harmed more by the disease than others. Horses are particularly susceptible. From information garnered from the Department of Ecology and Department of Health, it is anticipated that 30–40% of infected horses showing symptoms will die or be permanently disabled. Thus, it is imperative that horse owners have their horses vaccinated before the next mosquito (the vector) season begins (April). Why the urgency? It is because 2 doses, 3–6 weeks apart are required, and immunity is not achieved until 5 or 6 weeks after the second dose. This also requires an annual booster. Only a veterinarian can perform the vaccination.

Horse Lice

Lice feed on the blood of horses. Rarely, horses become anemic and run down. Horses also will rub excessively, wearing the hair coat in patches. Young lice hatch from yellow eggs glued to the horse's hair. They grow to adults in 2 or 3 weeks. Lice can be spread from one horse to another by currycombs, blankets, saddles, and direct contact.

Pests of Swine

Hog Louse

The hog louse is dull gray brown and 1/4 inch long. It pierces the hog's hide, removing blood through tubelike mouth parts.

The hog louse lives in clusters. During cold weather it appears especially inside the legs and ears, and the folds of the skin of the neck. But any part of the animal's body that offers protection may harbor the pest.

Eggs are laid throughout the winter and are glued to the hairs close to the skin. During this time of year, eggs take about 2 to 3 weeks to hatch, and the immature stage lasts 2 weeks. Immature stages and adults live entirely on the animals, except when they fall off accidentally.

Sarcoptic Mange or Itch Mite

Sarcoptic mange or itch is caused by a small (1/50 inch long) white or yellow mite that burrows into the hide. An animal having sarcoptic mange is unthrifty: its hide is rough and scaly, the hair stands erect, and it rubs against objects such as fence posts and corners of buildings.

The areas around the eyes, ears, and along the top of the back and neck are more often affected. These places may be scabby and covered with pimples. The problem occurs usually in the fall, winter, and spring.

The mange mite tunnels into the skin. The slender, winding tunnels of the mites are nearly 1 inch in length and occur throughout the infested part of the body. Eggs are laid in the tunnels; they hatch in 3 to 10 days. Under favorable conditions, the mites can complete a generation (brood) in 2 weeks.

Sarcoptic mange is highly contagious. The mites and their eggs can live in bedding and other places for weeks without food, making infestation or reinfestation of healthy animals easy.

Pest Management

A pest species can be any species that humans consider undesirable. Any organism that reduces the availability, quality, or value of a human resource can be classified as a pest. This designation in no way reflects the organism's role in the natural ecosystem but is more an indicator that they are in conflict with humans.

A pest in one area may not be considered a pest elsewhere. Often organisms rise to pest status because they escape normal control by natural regulating agents. This is achieved through direct or indirect importation to a new region or by human activities which reduce or eliminate the efficiency of their natural enemies. Without controls on population growth, organisms can rapidly achieve levels at which damage is caused thus becoming pests (e.g., locust swarms stripping landscapes bare). However, organisms do not need to exist in large numbers to be a pest. For example, the codling moth (Cydia pomonella) does not lay many eggs compared to many insects and often produces only one generation each year . However, because it blemishes apples, making them commercially undesirable, the codling moth is considered an important agricultural pest.

Pest management is therefore a means to reduce pest numbers to an acceptable threshold. An acceptable threshold, in most cases, refers to an economically justifiable threshold where application of pest control measures reduces pest numbers to a level below which additional applications would not be profitable (i.e., where additional costs of control exceed additional benefits). Pest eradication (i.e., complete removal) is usually not a viable option.

Methods of control can be categorized as chemical, biological, cultural, physical/mechanical, or genetic, and are discussed below.

- Chemical: Chemicals (e.g., insecticides, herbicides, rodenticides) can be broad-spectrum (non-selective) or narrow-spectrum (selective), and can be organic or inorganic. Chemicals used to regulate pest abundance can act as nerve toxins (for insects and mammals) and growth regulators/inhibitors. Chemicals can also be used to affect pest abundance through more indirect means, such as releasing pheromones to disrupt breeding behavior and interfere with mating. Chemical pesticides are often toxic to non-target organisms including the pest's natural enemies, can persist in the environment affecting water supply, soil productivity, and air quality, and can be biomagnified in the food chain. Inappropriate use of pesticides can result in target pest resurgence from killing off natural enemies, secondary pest outbreaks by removing natural enemies of other organisms and allowing them to rise to pest status, and evolved resistance to the pesticide.

- Biological: Due to any number of reasons, including those mentioned in the Chemical and Cultural sections, compromising the effectiveness of natural enemies often allows potential pest organisms to experience virtually unregulated population growth and enables them to reach pest status. Biological control involves the use of a pest's natural enemies (e.g., predators, pathogens, parasites and parasitoids), to control pest abundance. Measures to conserve or enhance the impact of natural enemies should be attempted first. Perhaps biological control is most known for importation of natural enemies, often from the pest's area of origin, to control non-native pests (e.g., importing vedalia bettles to control cottony cushion scales which were attacking California citrus orchards). A number of safeguards are necessary before implementing importation actions to ensure imported organisms will not pose additional threats to non-target organisms. A third approach to biological control involves augmenting natural enemies through rearing and periodic releases and can be inoculative (natural enemies are released early in the season) or innundative (natural enemies are released as a biological pesticide).
- Cultural: The effectiveness of natural enemies can be compromised by human practices. Application of broad-spectrum pesticides which kill off natural enemies in addition to target pest species, the type of crop plant, the crop environment, and cropping practices. Modern crop varieties often inadvertently create conditions which favor pest species (e.g., pest species which have bored deeper into larger fruit making them inaccessible to natural enemies). Crops are often monocultures, consisting of a single crop species, which creates a homogenous habitat often lacking key requirements of natural enemies, thus favoring pest species. Moreover, many harvesting practices prevent natural enemies from persisting in annual crops. Examples of cultural practices that encourage natural enemies and discourage pest persistence include intercropping (multiple crops in the same field) to make it more difficult for pests to find a host plant, planting trap crops which attract pests away from harvest crops and which can later be treated with select application of pesticides, and delaying planting times to coincide with times where pests have emerged and died off for the season.
- Physical: Manual or mechanical removal, or installation of physical barriers can be used to exclude pest species. Removal methods include use of animal traps, sticky cards for insects, manual removal of insects from plants (e.g., hand picking or spraying with a hose), removing diseased or infected materials (e.g., pruning branches or removing diseased litter). Physical barriers such as fences, nets, mulch, and tree trunk guards can exclude pests and reduce the damage they inflict.
- Genetic: Genetic alteration to reduce pest impacts is not as widely known or publicly available as other control options. Autocide is one type of genetic control and involves using

the pest itself to induce increased mortality rates. Sterile males are introduced into the population, which, after mating with females, creates infertile eggs. This is an expensive option with many limitations including potential for reduced competitive viability of the introduced sterile males versus naturally occurring fertile males. Straightforward genetic manipulation to create pest resistant plant strains is another form of controlling pest impacts. However, genetic manipulation research and development is costly, and introduces a whole other series of ethical and environmental issues that are not easily addressed. Genetic manipulation is not a viable control option for the general public.

Integrated Pest Management (IPM) is an increasingly popular process for controlling pests. IPM considers the ecosystem as a whole and takes into consideration a balanced mix of the aforementioned control methods to produce the most effective and least damaging plan. All the methods are mutually augmentative with chemical control means as the last resort in the plan. Ideally, an IPM plan would result in a sustainable system without need for much costly follow-up maintenance.

A number of insects and pathogens have been identified as pests particularly for their impacts to vineyards, orchards, and other agricultural industries important to Napa County's economy. The following table presents a number of important pest species in Napa County as well as their deleterious effects.

Pest Type Effects

Pierce's Disease (Xylella fastidiosa)	Fungus	A lethal disease of the grapevine which is often spread by insects with piercing/sucking mouthparts that feed on xylem sap such as the glassy-winged or blue-green sharpshooter.
Oak Root Fungus (Armillaria mellea)	Fungus	Parasitizes the roots of orchard trees, oaks and many other woody plants including grapevines sometimes causing sudden death.
Botrytis Bunch Rot (Botrytis cinerea)	Fungus	Directly affects grape bunches reducing yield and quality of grapes. Infected fruit causes off-flavors and aromas in wine, and is unsuitable for most wine production.
Powdery Mildew (Uncinula necator)	Fungus	Damages or destroys berries, reduces photosynthesis and can affect wine flavor and quality.
Grape Leafhopper (Erythroneura elegantula)	Insect	Damage vines by inserting their stylets to suck out the contents of leaf cells reducing the vine's photosynthetic ability, and are often a nuisance to vineyard workers during harvest
Spider Mites (Tetranychus pacificus, and Eotetranychus willamettei)	Insect	Destroys grapevine leaf tissue feeding on the lower surface of the leaf reducing photosynthesis and other physiological functions.
Pocket Gopher (Thomomys bottae)	Mammal	Damages vines by chewing on roots while tunneling underground, and above-ground portions as well. Gophers cause other problems by chewomg on plastic irrigation pipes, and building burrows which can divert and concentrate runoff, causing significant erosion problems in the vineyard as well as lawns and gardens.
Starling (Sturnus vulgaris)	Bird	Starlings, in addition to a number of birds, including house finches, and robins, flock consume ripening and mature grapes causing significant damage. In Napa County, flocks of starlings can reach numbers in the thousands, and damage from them can be the most severe.
Deer (Genus odocoileus)	Mammal	Damages vines by grazing on the foliage, often stripping canes clean of leaves causing extensive damage and stunting of vines.

References

- Plant-pests-and-diseases, emergency-types, emergencies: fao.org, Retrieved 1 May, 2019
- Locust-insect, animal: britannica.com, Retrieved 25 August, 2019
- Locusts, group, invertebrates, animals: nationalgeographic.com, Retrieved 31 July, 2019
- Japanese-beetle, beetles, insects, pests-and-problems, advice-tips-resources, help-for-the-home-gardener, your-garden, gardens-gardening: missouribotanicalgarden.org, Retrieved 4 June, 2019
- Corn-rootworms, insects, fieldcropsipm: purdue.edu, Retrieved 12 April, 2019
- Strange-life-potato-bug-jerusalem-cricket-child-earth: sciencetrends.com, Retrieved 15 May, 2019
- Pest-management: napawatersheds.org, Retrieved 21 July , 2019

Chapter 6

Crop Protection

The science and practice of managing plant diseases, weeds and pests that damage crops and forestry is termed as crop protection. It is concerned with weed management, plant disease management and insect management. This chapter has been carefully written to provide an easy understanding of the varied facets of crop protection.

Crop protection is the collection of tools, products, and practices farmers use to defend their crops against weeds, disease and harmful insects and fungi. These pests, which rob plants of water, sunlight and nutrients, can have a devastating impact on food production. Farmers around the world make multiple decisions each season on how best to protect their crops. Fortunately, advances in modern agriculture provide them with a variety of solutions.

Crop Protection Toolbox

There is no singular approach to crop protection. Farmers today have a diverse toolbox at their disposal to minimize damage from pests, including state-of-the-art chemical and biological products, advanced data analytics, and precision technologies. While these tools are individually powerful, combining them in an agronomic approach tailored to farmers' specific challenges enhances their effectiveness while minimizing the environmental impact of agriculture.

Crop Protection Innovations

As modern agriculture continues to evolve, so do our tools for managing weeds, insects, and disease. Crop management software, data science, precision farming technology, and soil health and nutrient management techniques are all ways technology is driving crop protection in modern agriculture. Chemical and biological crop protection products, nutrient management, and soil health practices are enhanced by data science, which transforms traditional practices into an integrated system focused on sustainable use. Real-time data drives the precise application of resources, and state-of-the-art software allows for quick, easy analysis of crop management and protection.

Crop Protection Chemicals

The development of chemicals to protect agricultural crops is an important activity within the chemical industry. Without them, many crops would suffer dramatic losses. Some of these chemicals, the insecticides, are also very important in combating human and animal diseases. The environmental and toxicological properties of these chemicals have improved considerably over the last six decades. Research aims to produce chemicals that are not just potent but are specific for the required purpose, whilst not affecting the environment in any other way. Because pests may develop resistance to crop protection chemicals there is a continual need for new products to be developed.

Three groups of chemicals dominate this part of the chemical industry. They are:

1. Herbicides: Substances that kill or inhibit growth of unwanted plants (weeds).
2. Insecticides: Substances that kill arthropod pests, i.e. insects and mites.
3. Fungicides: Substances that destroy or prevent the growth of pathogenic fungi. All three are pesticides.

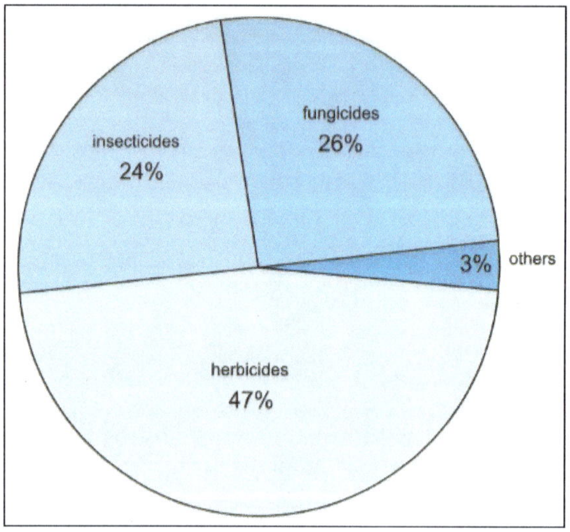

Global sales of crop protection chemicals (2008).

The effectiveness of a pesticide is the result of the proper 3D-assembly of specific groups in the chemical structure of its active ingredient. If several compounds of a given chemical class have related efficacies, they include a set of groups as a common minimum basic feature responsible for the best fit for the specific molecules in the biochemical target molecule (e.g. a protein) of the pest. This set of groups is called the toxophore. Where possible, the toxophores described in this unit are indicated by shading.

Herbicides

Herbicides are used to control the growth of unwanted plants (weeds). Modern herbicides generally act by restricting growth. They inhibit the action of one or more of the many receptors that catalyze reactions which are essential to the growth of the plant. There is one group however, the auxins, that kill by overstimulating growth. With selective herbicides, either the target in the weed is affected more than that of the crop, the herbicide is degraded more quickly within the crop, or the uptake or translocation of the active ingredient differs from that of weeds. Non-selective herbicides kill crops as well as weeds.

Herbicides can act in several ways:

Direct contact with plant tissues, for example, leaves; paraquat is a typical contact herbicide:

- By translocation (systemic herbicides),whereby the compound has the ability to be absorbed by aerial plant parts and is transported to roots (basipetal translocation); glyphosate and growth hormones belong to this group.
- By root uptake and transportation, to the upper parts of the plant (acropetal translocation).
- Through a combination of both methods; triketones are an example of this group as they are transported downwards and upwards.

Bipyridyliums

These are non-selective herbicides that act by interfering with photosynthesis. Paraquat is one of the most used:

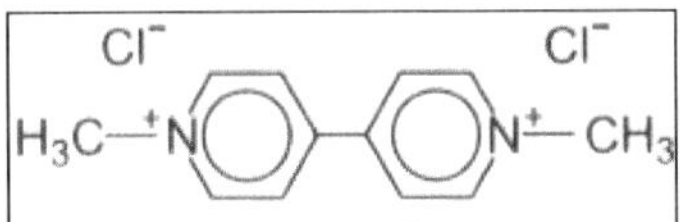

paraquat

Although extremely poisonous if swallowed, it binds tightly to soil and is quickly deactivated. It is sold under trade names such as Weedol and Gramoxone.

Auxins

Auxins are plant hormones. The most important member of the family is indole-3-ethanoic acid (indole-3-yl)acetic acid, IAA).

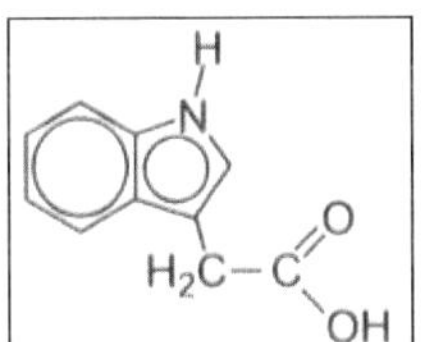

indole-3-ethanoic acid

They act by promoting growth of the plant but, given high doses, this can kill the plant and thus act as an herbicide. An example of a synthetic auxin is 2,4-dichlorophenoxyethanoic acid, often known as 2,4-D.

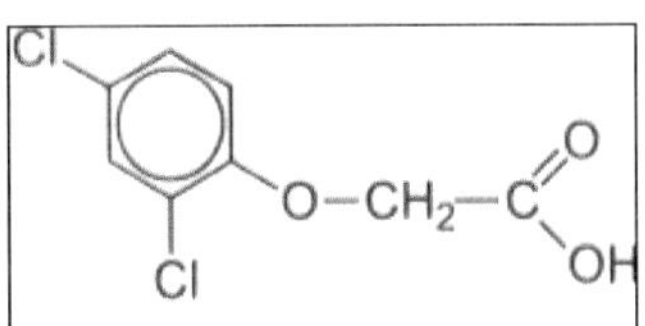

2,4-dichlorophenoxyethanoic acid

They are often used as one of the salts or esters. They are systemic herbicides and once absorbed act by translocation.

Two commercially sold auxins which mimic IAA are picloram and aminopyralid. They are both selective to broad leaved plants which absorb the herbicide through their leaves and translocate it to the roots. They can be used to kill weeds such as docks, thistles, dandelions and nettles amongst others.

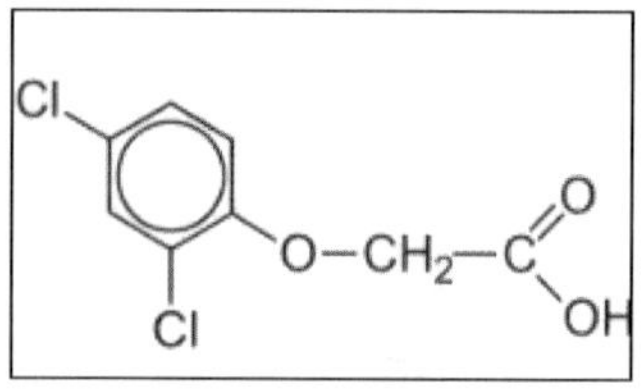

They are degraded by action of microorganisms in the soil and photochemically on the leaf of the plant.

Glycines

Glyphosate ((N-phosphonomethyl)glycine) is a non-selective systemic compound useful in the control of broad-leaved weeds and grasses. It is by far the most widely used herbicide with annual sales well in excess of £3 billion.

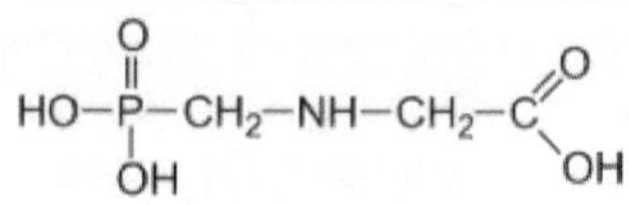

N-phosphonomethyl glycine (glyphosate)

The herbicide is absorbed by the leaves and then translocates to the roots. By inhibiting the action of an enzyme it prevents the production of aromatic amino acids needed by the plants for protein synthesis. As this enzyme is absent in mammals, glyphosate has little toxicity. It is very effective and is rapidly degraded in the soil and by light. It is marketed under many names including Roundup and Tumbleweed.

Although glyphosate is a non-selective herbicide, selective seeds of corn (maize) and soya have been developed (via genetic engineering) whose plants can withstand the herbicide. Thus, fields can be sprayed with the herbicide and these transgenic crops will be the only ones that will not be affected. These seeds are used in many parts of the world but have not been allowed in most EU countries, except in very tightly controlled trials because of public concern about the effect of transgenic crops on the environment.

There have been reports that some weeds have become resistant to glyphosate. However some crops, for example soya, are being developed which are also resistant to auxins such as 2,4-D and so farmers will be able to use a mixture of glyphosate and an auxin.

Glufosinate is another non-selective Herbicide:

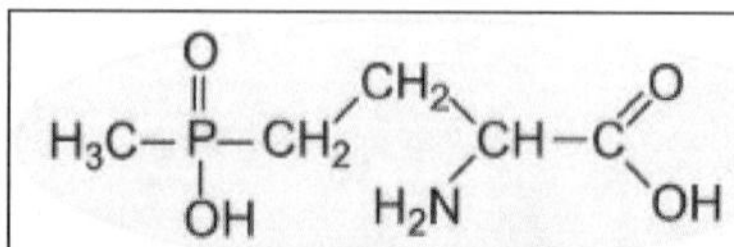

glufosinate

Glufosinate inhibits the formation of glutamine synthetase, an enzyme that forms the amino acid glutamine from glutamate, a process vital for photosynthesis. The weeds around these transgenic soybean plants in the US have been controlled by spraying the whole field with a solution of glufosinate, marketed as Ignite.

Sulfonylureas

Characteristic features of these chemicals are their potency on a wide range of weeds and their safety to mammals and the environment. They act in plants to block the biosynthesis of essential amino acids (valine and isoleucine), and thus prevent cell division and cell growth.

Mesosulfuron is another example of a herbicide which controls weeds in cereals and is marketed for example as Atlantis.

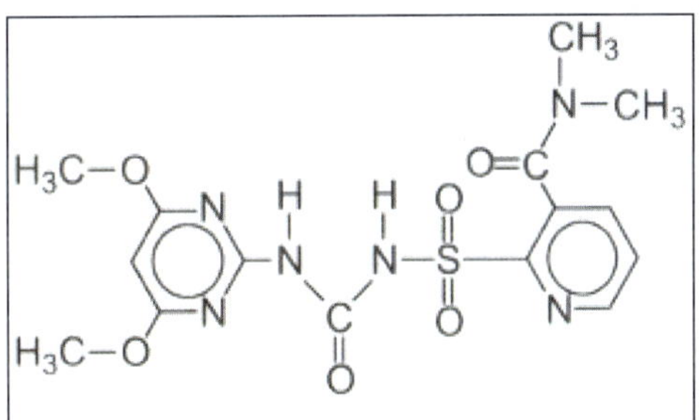

nicosulfuron

They are selective systemic herbicides which are absorbed by the foliage before translocating through the plant. They are particularly useful in controlling annual grass and dicot weeds in crops such as corn (maize) and are rapidly degraded in soils.

A well known example is nicosulfuron which is marketed, for example, as Accent which is used to control weeds in corn.

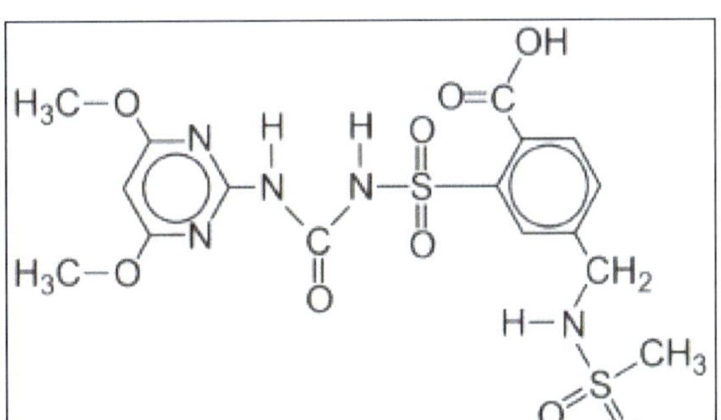

mesosulfuron

Triketones

The triketones act by inhibiting an enzyme that would have led to the formation of compounds needed in the biosynthesis of carotenoids. These, in turn, play a key role in absorbing light energy for photosynthesis and protecting chlorophyll from photochemical damage. Triketones act through application on the foliage or via the roots. In either case there is translocation through the plant. They are particularly effective in controlling a wide range of weeds. Degradation in soils is fairly rapid.

Tembotrione is an example that allows the control of grasses and broad leaves in corn and is sold under the trade name Laudis.

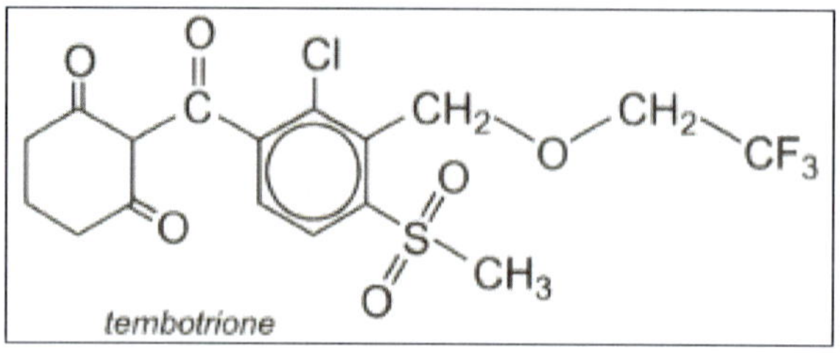

The weeds in this field of corn (maize) have been controlled with tembotrione. It has been applied in combination with a compound that triggers degradation of tembotrione in the corn but not in the weeds. Such compounds are known as safeners.

Mesotrione is another example of this class of herbicide and is sold under various trade names, including Callisto.

Inhibitors of Acetyl-coA Carboxylase

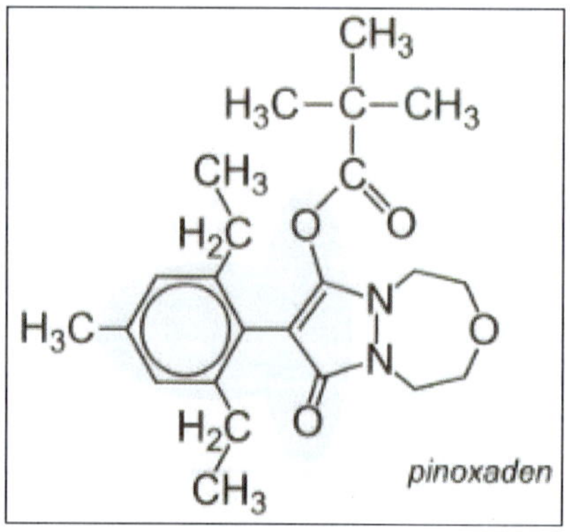

Some herbicides act by inhibiting the action of the enzyme, acetyl-CoA carboxylase, essential for the biosynthesis of carboxylic (fatty) acids, needed by the plant for the production of cell membranes. An example is pinoxaden which is particularly useful in dealing with grass in fields of barley and wheat.

The inhibitors are selective herbicides which are absorbed by the leaves of the plant and act following translocation. They are rapidly degraded in the soil.

Among the trade names for pinoxaden is Axial.

Insecticides

The world market for insecticides is dominated by compounds interfering with the nervous system of pest invertebrates, since this target organ usually provides rapid control. Insecticides acting on target sites such as acetylcholinesterase (organophosphates and methyl carbamates), voltage-gated sodium channels (pyrethroids), nicotinic acetylcholine receptors (neonicotinoids) and ligand-gated chloride channels (macrocyclic lactones and phenylpyrazoles) account for more than 75% of total insecticides sales.

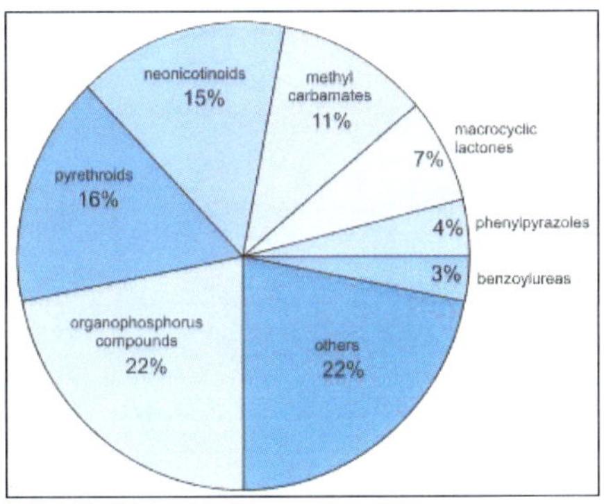

World market share of insecticides.

Other mechanisms such as interference with insect development are usually more selective (either by distinguishing between species or by their life cycles), but such compounds are generally much slower in their action.

Organophosphorus Compounds

First introduced in 1944, organophosphorus compounds (OPs) are economically still the most successful and diverse chemical class of insecticide. More than 100 different active ingredients belonging to this class are known. One of the most successful OPs is considered to be chlorpyrifos. This non-systemic insecticide, used on both the foliage and the soil, affects the stomach and respiratory action of the pest.

All OPs act by binding irreversibly to the enzyme, acetylcholinesterase (AChE). This prevents the hydrolysis of the neurotransmitter acetylcholine in the central nervous system (CNS) and leads to prolonged periods of nerve excitation. This then results in paralysis and subsequently death of the insects and their predators. OPs are used to control almost all pests including Lepidoptera (e.g. moths), Coleoptera (e.g. beetles), Diptera (e.g. flies, such as whiteflies, and mosquitoes), Hemiptera (e.g. aphids and leafhoppers). Additionally they control nematodes (parasitic worms) and mites. Four examples are:

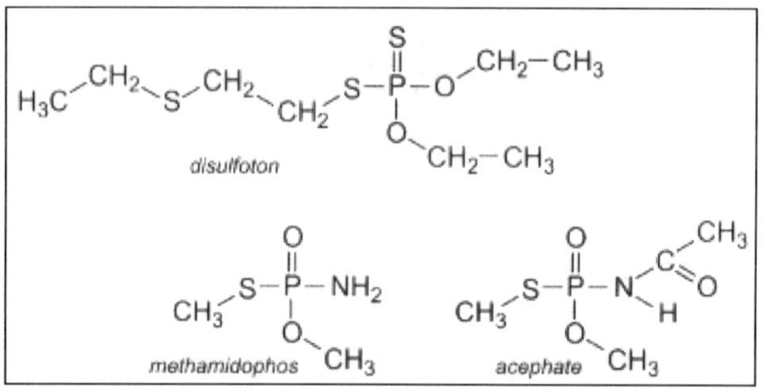

disulfoton, *methamidophos*, *acephate*

The major disadvantage of most OPs is their toxicity to vertebrates and this has resulted in the search for other compounds to replace them.

Organophosphorus compounds are rapidly broken down by enzymatic action in the soil.

Methyl Carbamates

Another important class of AChE inhibitors, introduced in the late 1950s, is the methyl carbamates which are usually less toxic to non-target organisms than the organophosphorus compounds.

Their structural features are found in a natural product isolated as early as 1864, physostigmin (eserin), the structure of which was finally elucidated in 1925.

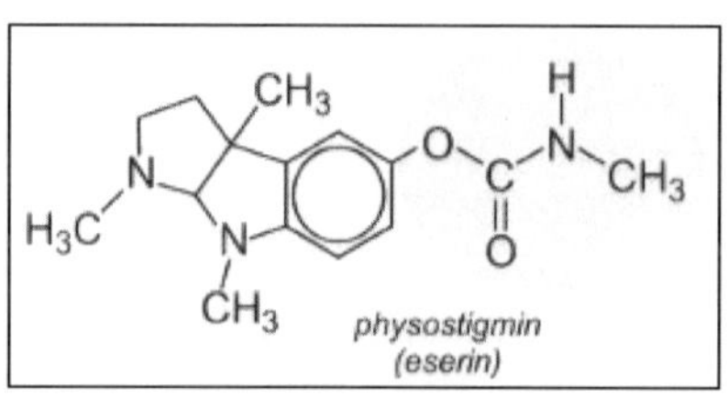

physostigmin (eserin)

They have short to moderate effective life, but are more selective in their effect. Pirimicarb is one of the most important methyl carbamates for aphid control. It was launched as a foliar spray for vegetables, arable crops and fruits in the early 1970s and is harmless to pollinating insects and aphid predators. Other widely applied carbamates are aldicarb which is used primarily as a soil insecticide and bendiocarb which is very important for the control of malaria-transmitting adult mosquitoes.

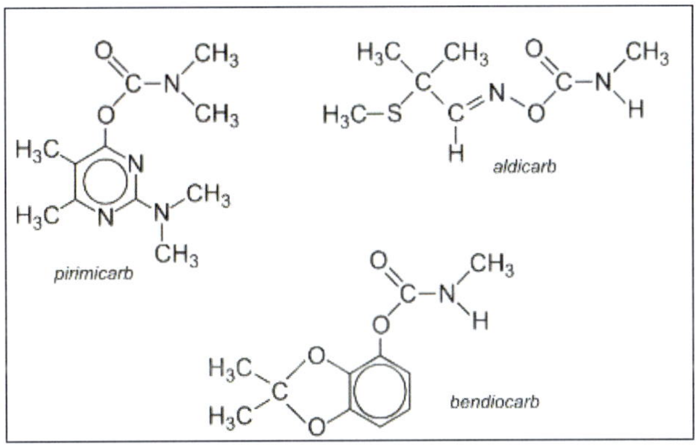

Macrocyclic Lactones (Avermectins and Milbemycins)

All natural and semi-synthetic 16-membered macrocyclic lactones disturb the systems which control the flow of chloride ions. This shuts down the electrical impulses in the nerve cells of their target organisms.

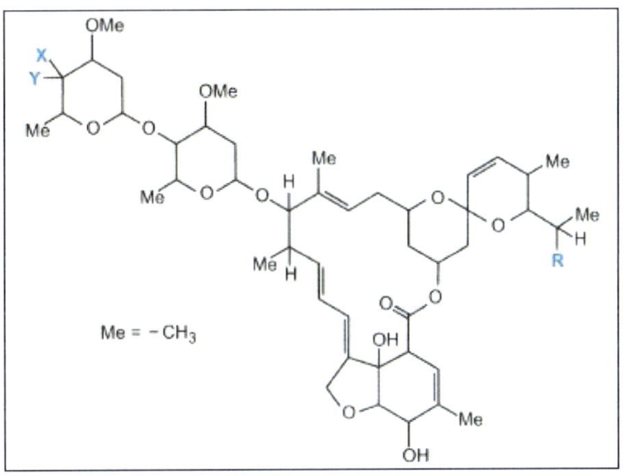

	X	Y	R
avermectin B1	OH	H	>80% C_2H_5 <20% CH_3
emamectin benzoate	H	NH-CH_3	>90% C_2H_5 <10% CH_3

The naturally occurring avermectins and emamectin benzoate are based on the above structure. The avermectins are produced by fermentation from the soil microorganism Actinomycetes (from the genus Streptomyces) and emamectin benzoate is prepared from abamectin in a series of chemical reactions. Another naturally occurring example is milbemycin.

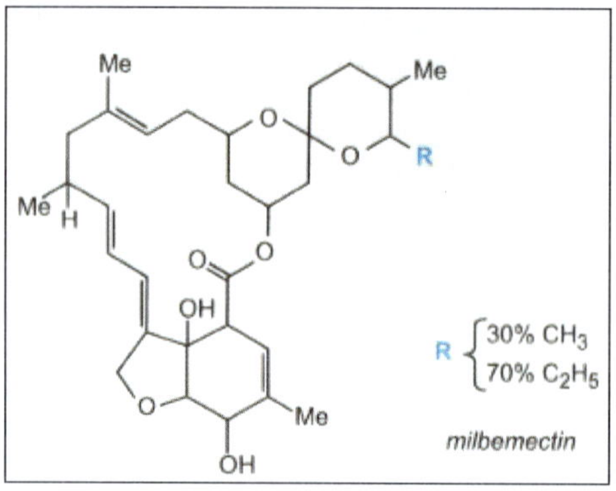

milbemectin

The whole family of macrocyclic lactones displays unprecedented potency against mites and insects as well as nematodes (e.g. parasitic worms). Avermectin is used in various crops such as citrus, pome fruits (for example, apples and pears), vegetables and cotton. Milbemectin is mainly used to combat the many different mites in tea and pome fruits (e.g.apples), and also against pine wood nematode which devastates pine trees in Japan and parts of the US.

The products are non-systemic and are removed rapidly from the environment after application. Photolysis on plant surfaces is fast, and they bind tightly to soil, where they are rapidly degraded by soil microorganisms. Thus, no leaching or bioaccumulation occurs. Because of the rapid uptake into sprayed foliage combined with fast degradation of surface residues, this family of insecticides is safe to use.

Phenylpyrazoles

A well-known member of this class of insecticides is fipronil, which can be applied to the foliage, soil and seeds. However, it has a limited ability to translocate through the plant. Fipronil acts by disturbing the chloride ion concentrations in cells in pest species such as the Lepidoptera (moths), Coleoptera (beetles) and Diptera (flies, mosquitoes). For example, it is widely used as a foliar spray for the control of leaf- and planthoppers in rice in south-east Asia.

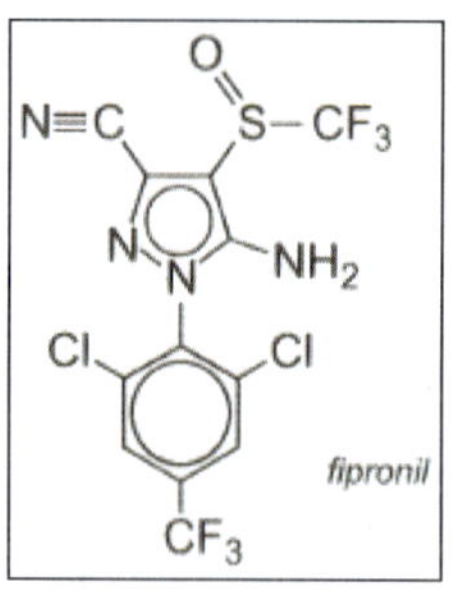

fipronil

Other applications of the phenylpyrazoles include the control of urban pests such as ants and cockroaches indoors and on lawns. They are also very effective in controlling termites. In animal health care, they are used on cats and dogs to combat ticks and fleas.

Nereistoxin Analogues and Neonicotinoids

Much work has been devoted to finding chemicals which will prevent the nicotinic acetylcholine receptor (nAChR) from functioning correctly. The receptor is important in allowing the passage of sodium and potassium ions in the nerve cells and thus affects the central nervous system. First examples of these insecticides are cartap, thiosultap, bensultap and thiocyclam.

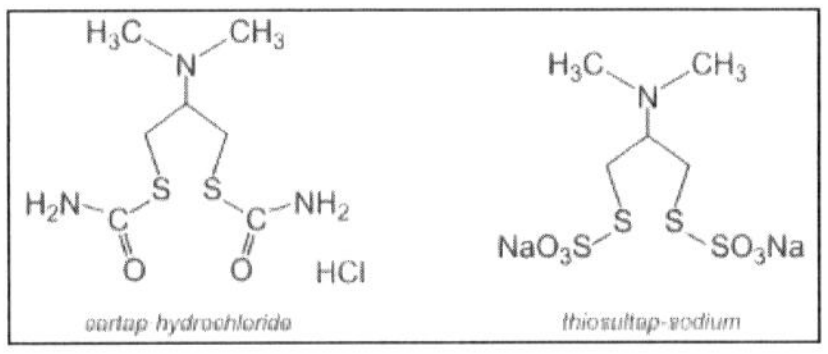

cartap hydrochloride   thiosultap-sodium

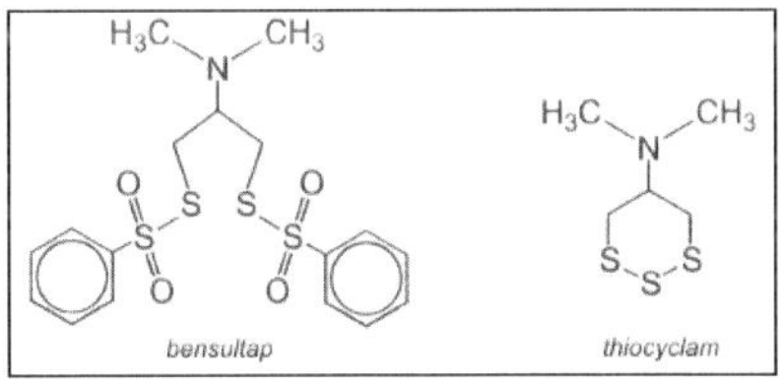

bensultap   thiocyclam

They break down either by action of water or light to produce the toxin, nereistoxin, a substance that was first isolated from the naturally occurring marine nereid worm Lumbrineris heteropoda Marenz, which acts by effectively paralyzing the pest.

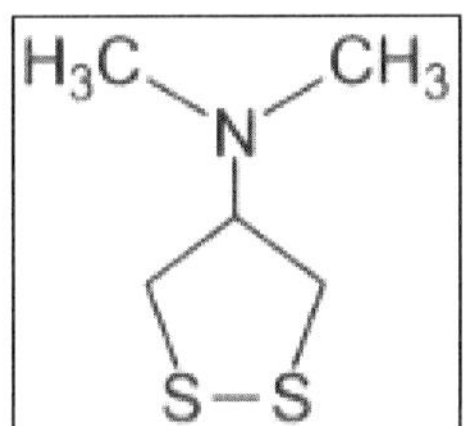

nereistoxin
(natural toxin)

Another group of insecticides which affect the nicotinic acetylcholine receptor is the neonicotinoids which are now the fastest growing and fourth major class of insecticides in crop protection. They are active against a broad range of insect pests, and exhibit activity through both oral (ingestion) and contact routes of application. They have a high level of efficacy, and a favourable environmental and toxicological profile. This has led to their rapid adoption in numerous agricultural areas for quick control of a broad range of chewing and sucking pests with minimum impact on beneficial insects.

Like the naturally occurring alkaloid nicotine (used for a long time in the form of aqueous tobacco extract),

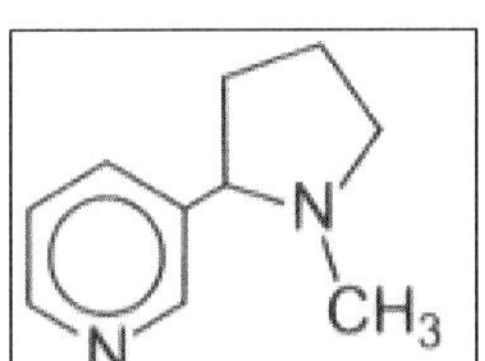

(s)-(-)-nicotine

Neonicotinoids act selectively on and overstimulate the insect's central nervous system. The following are examples of this class of pesticide.

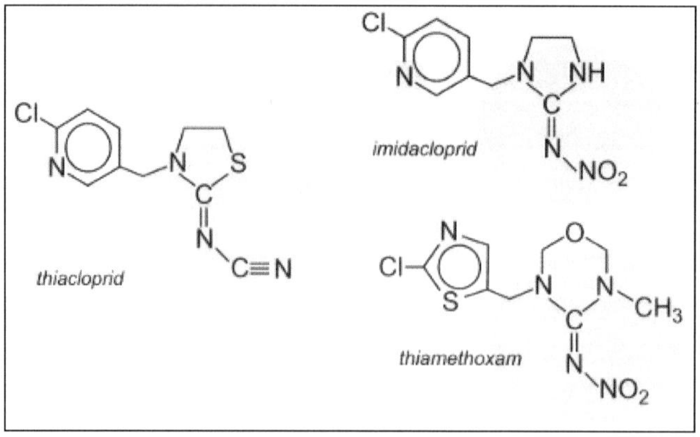

They have been shown to have long-lasting effect, both when treating the leaves and the seeds. Further, they have a relatively low risk for non-target organisms, although recently there have been concerns that they may be contributing to the significant decline in honey bee populations. The environment agencies in several countries are currently assessing the evidence for this.

Diamides

Diamides represent a new class of insecticidal chemistry, which include the phthalic acid diamides and the anthranilic diamides (chlorantraniliprole). They act on the cell's ability to control the flow of calcium ions (ryanodine receptors).

Flubendiamide is an example of a phthalic acid diamide:

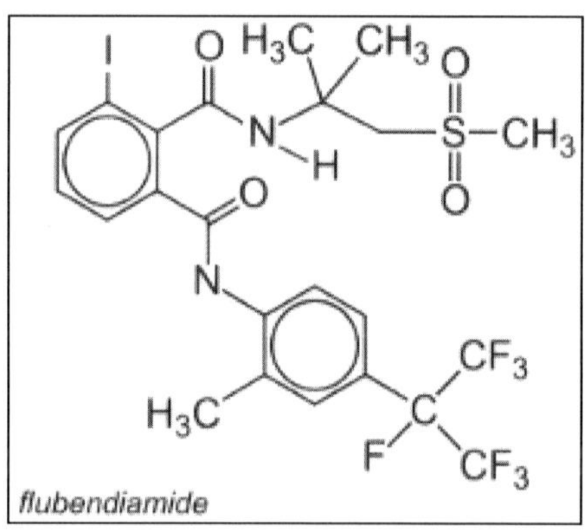

Chlorantraniprole is an example of an anthranilic diamide:

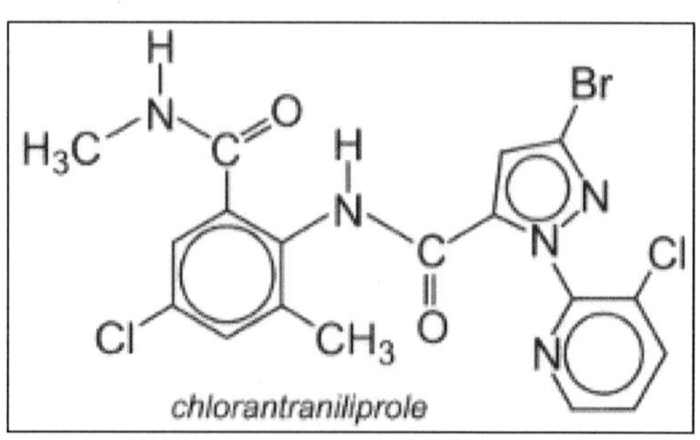

Rynaxypyr is used to protect cabbages from the ravages of caterpillars.

It is marketed under several names, for example Rynaxypyr.

The ryanodine receptor as an insecticide target site has been known for decades and is named after the natural plant-derived (alkaloid) insecticide, ryanodine. Although ryanodine's insecticidal properties are limited under field conditions, diamide insecticdes have been developed which are specific to insect ryanodine receptors and are particularly active against Lepidoptera (moths). Diamides protect fruit and vegetables from beetles, weevils, leaf miners and caterpillars.

They do not affect mammalian ryanodine receptors, most probably explaining their excellent toxicological profile, being specific to the pests and not affecting natural predators or pollinators and being non-toxic to mammals, fishes and birds.

Pyrethroids

Among the most important chemical classes of insecticides is the pyrethroids, which affect the flow of sodium ions in the nervous system. In the middle of the 19th century an insecticidal powder derived from dried flower heads of the genus Pyrethrum (Chrysanthemum) was introduced from Africa to central Europe. The insecticidal components were identified as pyrethrins. As they have several asymmetric centres these compounds have many stereochemical forms, and only a few of them are active as an insecticide. Natural pyrethrins I and II are unstable, sensitive to photo-degradation and relatively expensive. However, these natural pyrethrins were used as templates to generate analogues, the so-called synthetic pyrethroids. They act on contact with the plant.

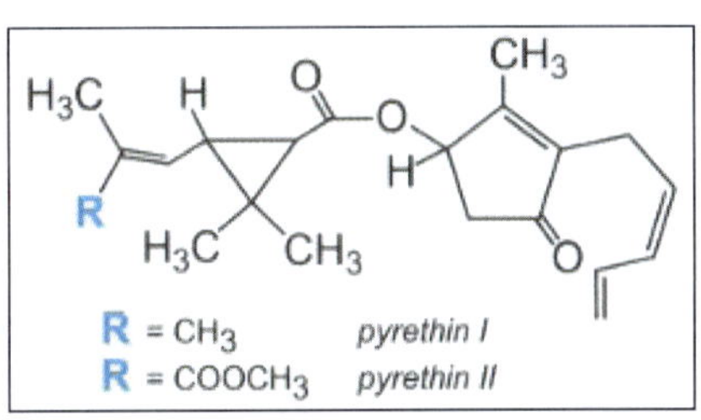

They can be separated into two classes according to the poisoning action they induce. Type I pyrethroids (e.g. permethrin) cause hyperactivity and incoordination, whereas Type II pyrethroids (e.g. deltamethrin) containing an alpha-cyano substituent induce nerve depolarization (polarization being needed to transmit the electrical signals in the nervous system) and subsequently paralysis of the insect. Pyrethroids are highly active against Lepidoptera pest species, but their speed of action, which leads to rapid knock-down, means they are used successfully in many crops against numerous pests including aphids.

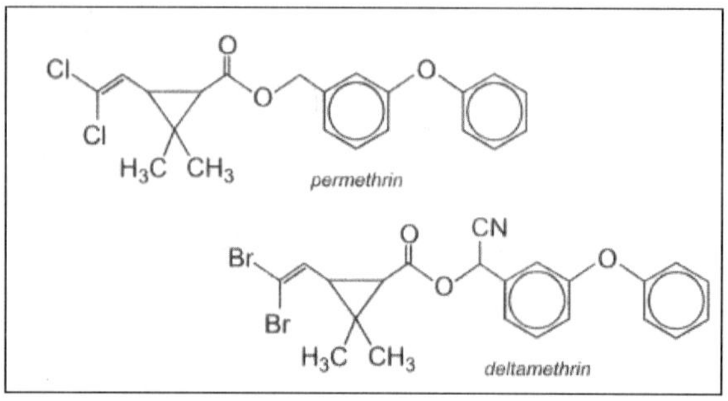

Pyrethroids are very useful insecticides as, although toxic to insects, they are harmless to mammals. They are non-systemic and are rapidly broken down in the soil. They are also used as indoor residual spray or on insecticide-treated nets for the control of malaria mosquitoes.

Benzoylureas

Benzoylureas have been developed and used as commercial insect growth regulators (IGRs) acting by inhibiting the biosynthesis of chitin (a long chain polymer of N-acetylglucosamine which is in the cell walls of the insect). Diflubenzuron was one of the first of this class to be used commercially when it was introduced in 1977. It is used for the control of chewing insects and coleopteran pests (beetles and weevils) in fruit, cotton, soybeans and vegetable crops.

Because of their non-toxicity to vertebrates, the benzoylureas such as lufenuron and triflumuron are also used in veterinary medicine and in the home against animal and human health pests such as fleas, ticks and cockroaches.

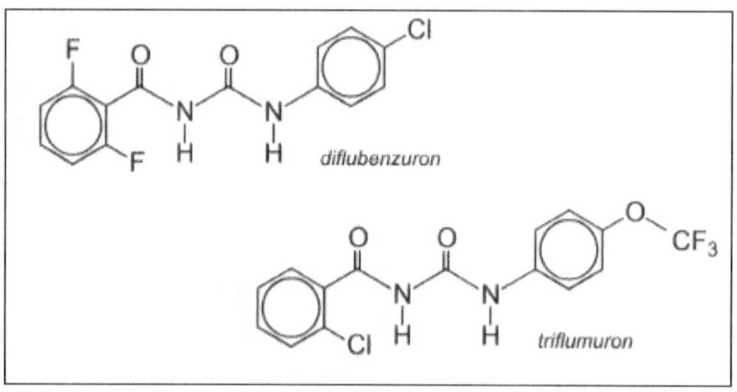

Ketoenols

Ketoenols are a new chemical class of insecticides affecting development by inhibiting acetyl CoA carboxylase and subsequent lipid biosynthesis. Spirodiclofen is a new non-systemic foliar acaricide (kills mites and ticks). It has particularly excellent long lasting efficacy and is effective in early to late season applications for spider mite control. Spirodiclofen is being developed for worldwide use in pome fruit (e.g apples, pears), stone fruit, citrus fruit, grapes, almonds and nuts, being very effective against mites. Spiromesifen is a new foliar contact acaricide and has been used worldwide on vegetables, fruits, cotton, corn, beans, tea and some ornamentals.

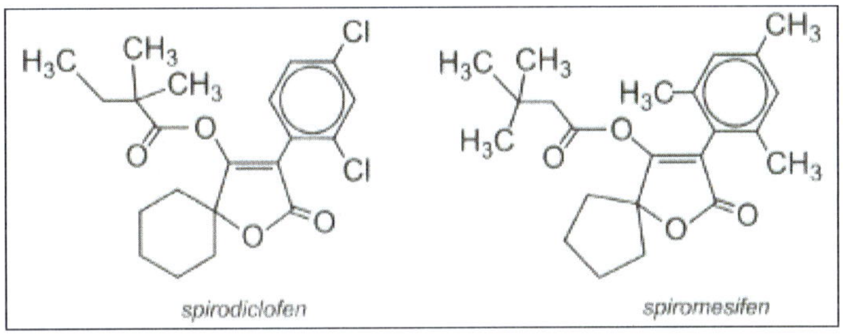

Fungicides

The term fungicide normally applies to synthetic chemical compounds that kill fungi or inhibit their growth. However, certain biological organisms can also be used to control fungal infections, which include mildews, rusts and leaf spots.

Some fungicides offer protection against the development of fungi (termed protective), and others cure the plant by eliminating the fungus (termed curative).

Nearly all modern fungicides (except for inorganic salts, see below) degrade in aerobic soils and therefore they are environmentally friendly, the rate of degradation depending on the temperature and humidity.

Some problem fungi on important crops. From left to right: potatop blight in its early stages; apple scab; grey mould on grapes.

Since fungi rapidly develop resistance to these chemicals, it is necessary to be able to draw on fungicides from different chemical classes and with different sites of action. The compounds discussed here are representatives of each important chemical class but the list is not exhaustive.

Triazoles

The 1,2,4-triazoles are one of the most important classes of fungicide.

They offer protection against a wide range of fungi that occur in cereals, rice, beet, trees, vegetables and flowers. These include amongst others powdery mildews, leaf spot diseases and rusts. The triazoles inhibit important biochemical pathways that produce sterols which are essential components of cell membranes.

The compounds are applied to the leaf surface of the plant and are distributed all over the plant (systemic activity). They act by both preventing the action of the fungi by inhibiting the germination of the spores and also curing the problems caused by the fungi. They are thus both protective and curative.

Examples of commercially available triazoles include:

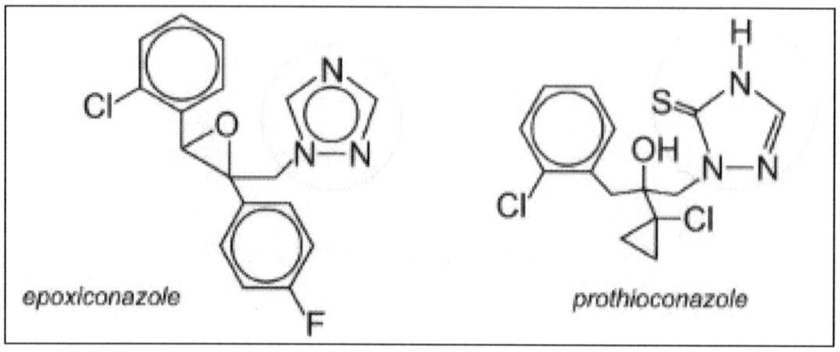

Strobilurins

The development of a chemical class called strobilurins was inspired by the study of a group of naturally occurring fungicides, for example, strobilurin A.

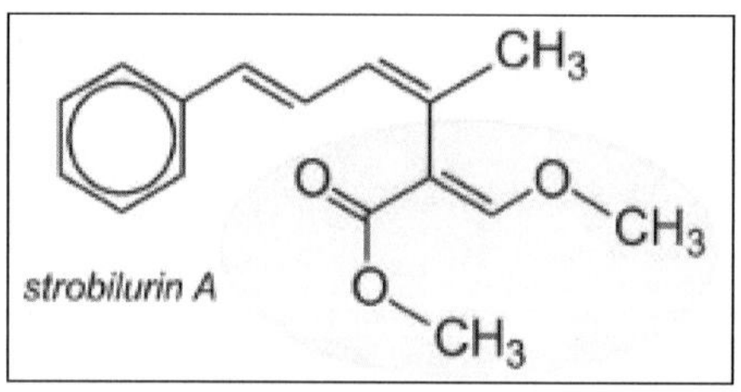

Strobilurins interfere with the production of adenosine-5'-triphosphate (ATP), the nucleotide that transports energy within cells for metabolism, thus preventing germination and growth. The strobilurins can combat most major fungal diseases found on grasses (turf), vines, fruit and particularly cereals.

Like the triazoles, they are systemic, and they offer protection and also cure the problems caused by the fungi.

Examples of commercially available strobilurins are given below:

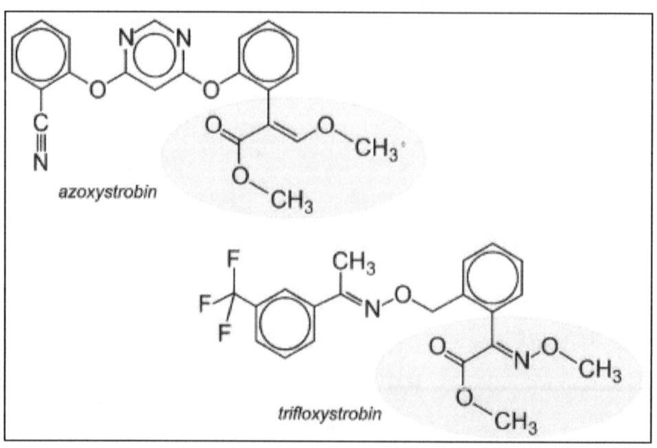

Carboxamides

Carboxamides are so named as they contain the carboxylic acid amide (-CO-NH2) or related group. As with the strobilurins, they interfere with the production of ATP.

An example of a carboxamide fungicide is boscalid used in particular to control powdery mildew on fruit and vegetables.

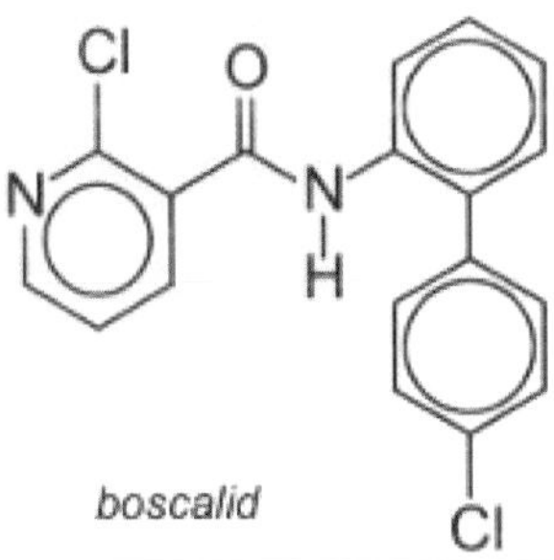

boscalid

It is expected that several other compounds in this class will be commercially available in the near future examples of which are bixafen and isopyrazam.

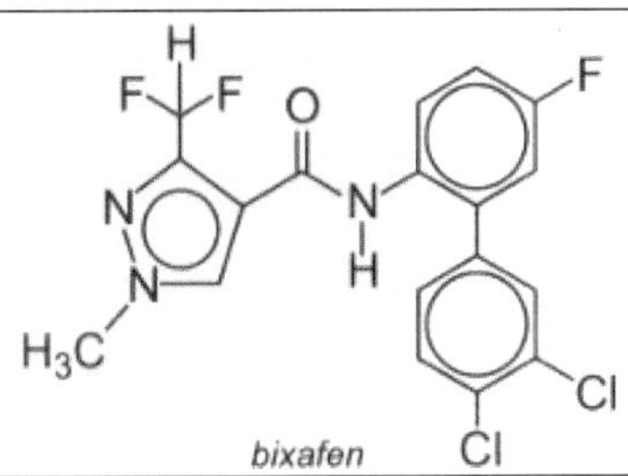

bixafen

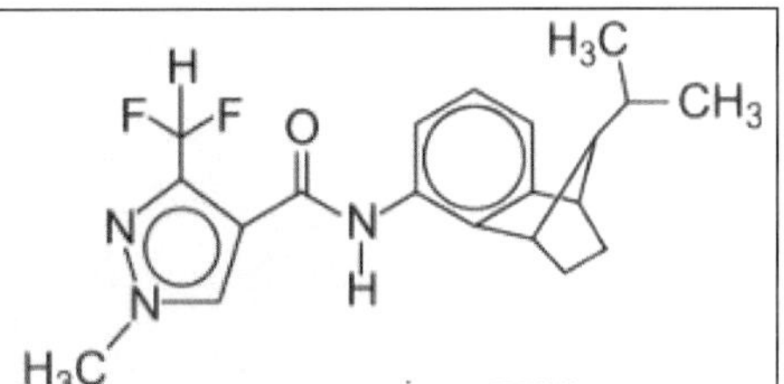

isopyrazam

Fungicides with a 'Multi Site Action'

Besides fungicides which act mainly on one target enzyme, several fungicides with a 'multi site action' are commercially available. They act unspecifically on a number of enzymes and are therefore not very sensitive to resistance problems. They are all protectant fungicides, acting on contact with the fungal spores on the surface of the leaves.

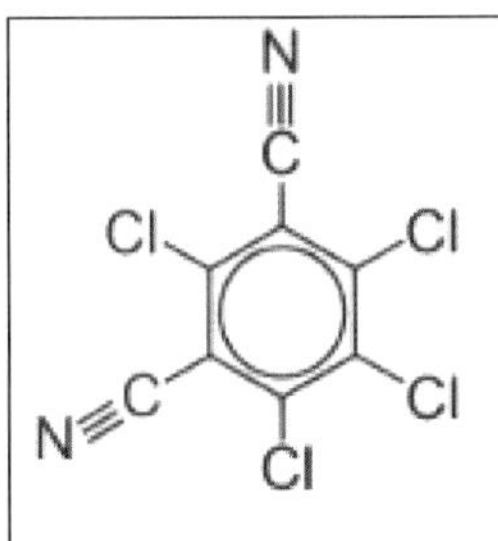

chlorothalonil

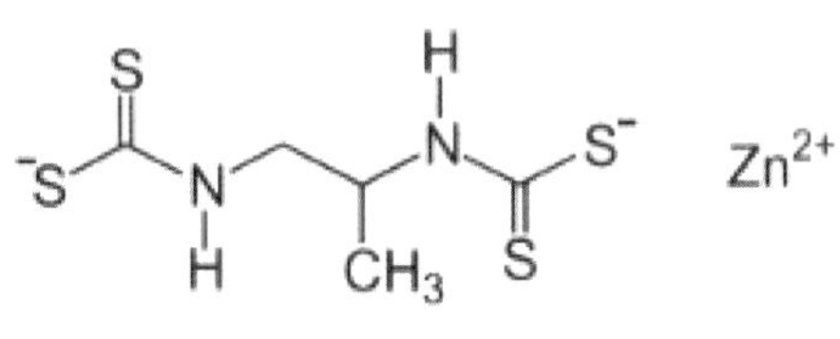

propineb

Some examples are inorganic compounds such as elemental sulfur and some copper salts of which the 'Bordeaux mixture' (a mixture of copper(II) sulfate and calcium hydroxide) is the most famous. Both are allowed to be used in the production of organic food. Examples of organic compounds with a multi site action are salts of dithiocarbamates such as chlorothalonil and propineb. They are very widely used on a wide range of plant pathogens in fruits and vegetables such as grapes, apples, tomatoes and potatoes.

Specific Fungicides with Activity Against Downy Mildews

Since downy mildews are a special class of plant pathogens causing severe damage mainly in grapes and potatoes a number of specific fungicides active only against this kind of plant diseases are available. Most of them act systemically. Since they belong to different chemical classes they exhibit several modes of action.

Examples are metalaxyl and iprovalicarb:

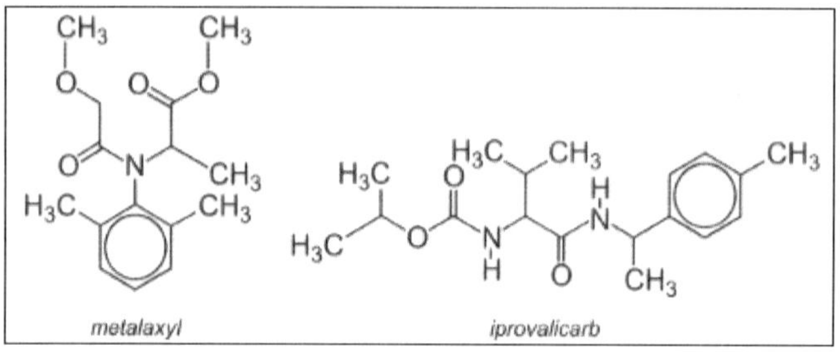

Crop Protection Methods

Weed Management

Weed is general term for any plant growing where it is not wanted. Ever since humans first attempted the cultivation of plants, they have had to fight the invasion by weeds into areas chosen for crops. Some unwanted plants later were found to have virtues not originally suspected and so were removed from the category of weeds and taken under cultivation. Other cultivated plants, when transplanted to new climates, escaped cultivation and became weeds or invasive species. The category of weeds thus is ever changing, and the term is a relative one.

Weeds interfere with a variety of human activities, and many methods have been developed to suppress or eliminate them. These methods vary with the nature of the weed itself, the means at hand for disposal, and the relation of the method to the environment. Usually for financial and ecological reasons, methods used on a golf course or a public park cannot be applied on rangeland or in the forest. Herbicide chemicals sprayed on a roadside to eliminate unsightly weeds that constitute a fire or traffic hazard are not proper for use on cropland. Mulching, which is used to suppress weeds in a home garden, is not feasible on large farms. Weed control, in any event, has become a highly specialized activity. Universities and agricultural colleges teach courses in weed control, and industry provides the necessary technology. In agriculture, weed control is essential for maintaining high levels of crop production.

The many reasons for controlling weeds become more complex with the increasing development of technology. Plants become weeds as a function of time and place. Tall weeds on roadsides presumably were not problematic prior to the invention of the automobile. However, with cars and increasing numbers of drivers on roads, tall weeds became dangerous, potentially obscuring drivers' visibility, particularly at intersections. Sharp-edged grasses are nominal nuisances in a cow pasture; when the area is converted to a golf course or a public park, they become an actual nuisance. Poison oak (Toxicodendron diversilobum) is rather a pleasant shrub on a sunny hillside in the open country; in a camp ground it is a definite health hazard. Such examples could be given ad infinitum to cover every aspect of agriculture, forestry, highway, waterway and public land management, arboretum, park and golf-course care, and home landscape maintenance.

Weeds compete with crop plants for water, light, and nutrients. Weeds of rangelands and pastures may be unpalatable to animals, or even poisonous; they may cause injuries, as with lodging of foxtails (Alopecurus species) in horses' mouths; they may lower values of animal products, as in the cases of cockleburs (Xanthium species) in wool; they may add to the burden of animal care, as when horses graze in sticky tarweeds (Madia species). Many weeds are hosts of plant disease organisms. Examples are prickly lettuce (Lactuca scariola) and sow thistle (Sonchus species) that serve as hosts for downy mildew; wild mustards (Brassica species) that host clubroot of cabbage; and saltbrush (Atriplex species) and Russian thistle, in which curly top virus overwinters, to be carried to sugar beets by leafhoppers. Many weeds are hosts of insect pests, and a number are invasive species.

Modern weed control can be classified as mechanical, chemical, or biological.

Mechanical Control

Mechanical weed control began when humans first pulled weeds from their cereal crops and attempted to grow single plant species, free from all plant competition. That was the start of monoculture, a method that since has come to dominate agriculture. Contrary to the principles of ecology, farmers throughout the world grow the major food, fibre, and forage crops in a monoculture because experience has shown that the highly improved modern crop species give their highest yield under this system.

Peanut farming.

From hand pulling, humans devised simple tools such as the spud, the knife, and the hoe to eliminate weeds. For thousands of years, from the Egyptian culture to the Renaissance, those simple methods were used. The first efforts to turn away from simple hand methods and mechanize the arduous task of weed control began in 17th-century England. Since then there has been continuous improvement of agricultural tools used to destroy weeds and of cultural methods employed to minimize weed growth. The principal virtue of cultivation of row crops is the control of weeds. Any method of weed control that minimizes tillage tends to conserve soil structure and maintain fertility.

In addition to tillage, other mechanical methods of weed control involve burning, grazing, and the use of ducks or geese in certain crops (in cotton and mint especially). All of those methods have drawbacks: there is the arduous, painful nature of hand weeding; the repetitious and often harmful nature of clean tillage with machinery; the slow, fuel-consuming nature of burning; and the costly requirement of livestock or fowl for the biological grazing methods. Tillage, still the most widely used method of row-crop weed control, has been greatly improved by development of precision seeding and close preadjustment of tiller tools, allowing the passage of weed knives within an inch or less of the young crop plants. Despite these improvements it is known that weed knives injure crop roots, especially late in the tillage season. Additionally, tillage tools can spread perennial weeds rapidly, bringing about rapid infestation of whole fields.

Tillage.

Such methods as crop rotation, use of smother crops, use of weedfree seed, mulching and covering, and cleaning of machinery to prevent spread of weed seeds are also classified as mechanical.

Chemical Control

Chemical weed control has been used for a very long time: sea salt, industrial by-products, and oils were first employed. Selective control of broad-leaved weeds in fields of cereal crops was discovered in France in the late 1800s, and this practice soon spread throughout Europe. Sulfates and nitrates of copper and iron were used; sulfuric acid proved even more effective. Application was by spraying. Soon sodium arsenite became popular both as a spray and as a soil sterilant. On thousands of miles of railroad right-of-way, and in sugarcane and rubber plantations in the tropics, the hazardous material was used in tremendous quantities, often resulting in the poisoning of animals and occasionally humans. Diesel oil, as a general herbicide, and sodium dinitrocresylate (Sinox), as a selective plant killer, were introduced during the first three decades of the 20th century.

Sinox, the first major organic chemical herbicide, was developed in France in 1896. In the late 1940s new herbicides were developed out of the research during World War II, and the era of the miracle weed killers began. Within 20 years over 100 new chemicals were synthesized, developed, and put into use. Chemical weed control superseded both plant-disease and insect-pest control in economic impact. The year 1945 marked the beginning of a new era in chemical weed control. Introduced then were 2,4-D (2,4-dichlorophenoxyacetic acid), 2,4,5-T (2,4,5-trichlorophenoxy-acetic acid), and IPC (isopropyl-N-phenylcarbamate), the first two selective as foliar sprays against broad-leaved weeds, the third selective against grass species when applied through the soil. The new herbicides were revolutionary in that their high toxicity allowed for effective weed control at dosage rates as low as one to two pounds per acre. This contrasted with carbon bisulfide, borax, and arsenic trioxide, which were required at rates of up to one ton per acre, and with sodium chlorate, required at rates of around 100 pounds per acre. However, some of those early herbicides, including 2,4,5-T, were later deemed unsafe for humans and the environment and were discontinued in many countries. Effective herbicides have continued to be developed, and some, such as glyphosate, are widely used around the world.

Herbicide.

Herbicides may be grouped into two categories: selective and nonselective. Each category may be subdivided into foliage-applied and soil-applied materials and, in cases where field crops are treated, the application may be made before sowing the crop (preplanting), after sowing but before emergence of seedlings (pre-emergence), or after seedlings have emerged (postemergence).

A great advantage of chemical over mechanical weed control is the ease of application. This is particularly true in cereal croplands, pastures, rangelands, forests, and other situations where an airplane can be used. Many millions of acres are treated from the air each year, and modern equipment for treating row-crop land with herbicides has made weed control increasingly convenient. Sprayers, soil incorporation equipment, and spreaders for pelleted herbicides have all added to the convenience of, and removed uncertainty from, herbicide application. Machinery is available that simultaneously builds up beds, plants the seed, sprays with insecticide, and incorporates fertilizer and pre-emergence herbicide all in one operation.

Certain agricultural plants, known as herbicide-resistant crops (HRCs), have been genetically engineered for resistance to specific chemical herbicides, notably glyphosate. These genetically modified organisms (GMOs) have been available since the mid-1980s and enable effective chemical control of weeds, since only the HRC plants can survive in fields treated with the corresponding herbicide. However, because these crops encourage increased application of chemicals to the soil rather than decreased application, they remain controversial with regard to their environmental impact and general safety.

Genetically modified (GM) barley: The GM barley was investigated for its effects on soil quality.

Biological Control

Efforts to eradicate weeds through biological control are a fairly recent development. An early report from 1902 described the importation of insects from Mexico to Hawaii in an effort to control Lantana, an imported shrubby climbing weed that had spread over thousands of acres of pastureland, rendering them useless for grazing. Prickly pear cacti have been very effectively controlled in Australia; some 24 million hectares (60 million acres) have been converted from cactus thicket to plowland and pasture by the cactus moth (Cactoblastis cactorum), which was introduced from Argentina in 1925. By 1933 the major cactus areas were under control.

Another successful use of biological weed control was in the western United States, where Saint-John's-wort, or Klamath weed (Hypericum perforatum), was subjected to depredation by three insect species, beginning in California in 1945. The release of two insects of the genus Chrysolina and one of the genus Agrilus continued for a number of years, and the effort was carried to Oregon, Washington, and Idaho by 1950. The insects spread rapidly after introduction and became established, ultimately reclaiming some 1 million hectares (2.5 million acres) of rangeland from the invasive plant.

A number of vertebrate animals have been used to control certain specific weeds. Sheep and goats have been employed to control brushy plants on rangelands in many countries. Their effectiveness is evident in parts of the Middle East and Africa where dry range and desert lands have been almost completely denuded by grazing goats. In these cases, however, the destructiveness of the goats far outweighs their usefulness in plant pest control, which indicates the need for rational management in all efforts at weed control.

Geese have been used to control weeds in cotton fields in California and in mint plantations in Oregon. Certain fishes are useful in keeping aquatic plants under control; examples are the Congo tilapia, the Israeli carp, and the grass carp. The Florida manatee is known to consume many aquatic plants, and the giant ramshorn snail (Marisa cornuarietis) feeds on alligator weed, pondweed, and water hyacinth.

Although much more desirable than chemical weed control from the standpoint of longtime ecological effects, biological weed control has definite limitations. It is ideal for situations where a weedy plant was introduced free of its natural predators, and there are places where it offers the only hope for coping with serious weed situations. However, biological control using insects is limited almost entirely to perennial plants and is restricted mainly to weeds of uncultivated areas. Given the broad spectrum of agricultural weed species, the great number of seeds in the soil seed bank, and the fact that many weed seeds can survive for decades in the soil, success against annual crop weeds with biological agents is not very feasible. In those cases cheap and effective herbicides have proved most useful.

Additionally, biological weed control tends to be only periodically effective. Weed populations may be reduced initially to very low levels by the control insects; however, they recover when the insects die off for lack of food. The insect predators may then flourish again in reciprocal cycles, but complete control may never be achieved.

Finally, the introduction of alien organisms as biological control is hazardous in that these same organisms may become pests in the new habitat. Kikuyu grass, which was introduced into California to prevent soil erosion on hillsides and roadways, soon spread into orchards, turf, and crop areas, where it became a serious weed.

Cultural Control

Cultural control is usually associated with farming systems, although some elements are relevant to landscape and bushcare practices. It largely involves manipulating farming practices to suppress weed growth and production, while promoting the development of the desired plant.

The principles and techniques used to prevent weed spread are relevant to cultural control methods to limit the spread of weeds between different land areas.

Cultural Control Methods

Encourage the competitiveness of desired species that are more competitive and fast growing. This suppresses weed growth by reducing access to available sunlight, nutrients and moisture and can include:

- Choose plant and crop species or cultivars that are naturally more competitive. This can include using plant species that suppress other plant species by the release of toxins.
- Use high quality (large and plump) seeds, as they are more likely to produce vigorous and competitive plants.
- Use increased seeding rates and narrow row spacing.
- Use shallow seeding techniques, where possible, to allow the desired species to grow above the soil surface more quickly.
- Ensure the desired plant is placed in the optimum growing environment.
- Use fertilisers in the optimal growth period to encourage rapid growth of the desired species.
- If possible use plant species that are native to the local environment.

Make it hard for weeds to adapt to weed management techniques. Using the same land management routines year after year may result in weeds adapting to these practices. Some practices that make it hard for weeds to adapt and therefore reduce their spread and vigour include.

Rotate crops: if a weed has adapted to grain crops continuously being sown, then alternating with a broadleaf crop will remove the environmental condition to which the weed has adapted.

Rotate species with different seasonal and growing cycles.

Rotate herbicides with different modes of action to help delay the development of herbicide resistance.

Insect Management

Insects injure plants by chewing leaves, stems and roots, sucking juices, egg laying or transmitting diseases.

Injury by Chewing Insects

Insects take their food in a variety of ways. One method is by chewing off external plant parts. Such insects are called chewing insects. It is easy to see examples of this injury. Perhaps the best way to gain an idea of the prevalence of this type of insect damage is to try to find leaves of plants with no sign of insect chewing injury. Cabbageworms, armyworms, grasshoppers, the Colorado potato beetle and the fall webworm are common examples of insects that cause chewing injury.

Injury by Piercing-Sucking Insects

Another important method which insects use to feed on plants is piercing the epidermis (skin) and sucking sap from cells. In this case, only internal and liquid portions of the plant are swallowed, while the insect feeds externally on the plant. These insects have a slender and sharp pointed part of the mouthpart which is thrust into the plant and through which sap is sucked. This results in a very different but nonetheless severe injury. The hole made in this way is so small that it cannot be seen with the unaided eye, but the withdrawal of the sap results in either minute white, brown or red spotting on leaves, fruits and/or twigs; leaf curling; deformed fruit; or a general wilting, browning and dying of the entire plant. Aphids, scale insects, squash bugs, leafhoppers and plant bugs are examples of piercing-sucking insects.

Injury by Internal Feeders

Many insects feed within plant tissue during a part or all of their destructive stages. They gain entrance to plants either in the egg stage when the female thrust into the tissues with sharp ovipositors and deposit the eggs there, or by eating their way in after they hatch from the eggs. In either case, the hole by which they enter is almost always minute and often invisible. A large hole in a fruit, seed, nut, twig or trunk generally indicates where the insect has come out, and not the point where it entered.

The chief groups of internal feeders are indicated by their common group names: borers; worms or weevils in fruits, nuts or seeds; leaf miners; and gall insects. Each group, except the third, contains some of the foremost insect pests of the world. In nearly all of them, the insect lives inside the plant during only a part of its life and emerging sooner or later as an adult. Control measures for internal feeding insects are most effective if aimed at adults or the immature stages prior to their entrance into the plant.

A number of internal feeders are small enough to find comfortable quarters and an abundance of food between the upper and lower epidermis of a leaf. These are known as leaf miners.

Gall insects sting plants and cause them to produce a structure of deformed tissue. The insect then finds shelter and abundant food inside this plant growth. Although the gall is entirely plant tissue, the insect controls and directs the form and shape it takes as it grows.

Injury by Subterranean Insects

Subterranean insects are those insects that attack plants below the surface of the soil. They include chewers, sap suckers, root borers and gall insects. The attacks differ from the above ground forms only in their position with reference to the soil surface. Some subterranean insects spend their entire life cycle below ground. In other subterranean insects, there is at least one life stage that occurs above the soil surface; these include wireworm, root maggot, pillbug, strawberry root weevil, and corn rootworm. The larvae are root feeders while the adults live above ground.

Injury by Laying Eggs

Probably 95% or more of insect injury to plants is caused by feeding in the various ways just described. In addition, insects may damage plants by laying eggs in critical plant tissues. As soon as the young hatch, they desert the plant causing no further injury.

Use of Plants for Nest Materials

In addition to laying eggs in plants, insects sometimes remove parts of plants for the construction of nests or for provisioning nests.

Insects as Disseminators of Plant Diseases

Insects may spread plant diseases in the following ways:

- By feeding, laying eggs or boring into plants, they create an entrance point for a disease that is not actually transported by them.
- They carry and disseminate the causative agents of the disease on or in their bodies from one plant to a susceptible surface of another plant.
- They carry pathogens on the outside or inside of their bodies and inject plants hypodermically as they feed.
- The insect may serve as an essential host for some part of the pathogens life cycle, and the disease could not complete its life cycle without the insect host.

Examples of insect vectored plant diseases are shown below:

Disease	Vector
Fireblight (bacterial)	Pollinating Insects
Tomato curly top (virus)	Beet leafhopper
Cucumber mosaic (virus)	Aphids

Benefits and Value of Insects

Insects must be studied carefully to distinguish the beneficial from the harmful. Producers have often gone to great trouble and expense to destroy insects, only to learn later that the insect destroyed was not only harmless, but it was actually engaged in saving their crops by eating destructive insects.

Insects are beneficial to the vegetable grower in several ways:

- Insects aid in the production of vegetables by pollinating the blossoms. Melons, squash and many other vegetables require insects to carry their pollen before fruit set.
- Parasitic insects destroy other injurious insects by living on or in their bodies and their eggs. Insects also act as predators, capturing and devouring other insects.
- Insects destroy various weeds in the same ways that they injure crop plants.
- Insects improve the physical condition of the soil and promote its fertility by burrowing throughout the surface layer. Also, the dead bodies and droppings of insects serve as fertilizer.
- Insects perform a valuable service as scavengers by devouring the bodies of dead animals and plants and by burying carcasses and dung.

Many of the benefits from insects enumerated above, although genuine, are insignificant compared with the good that insects do fighting among themselves. There is no doubt that the greatest single factor in keeping plant-feeding insects from overwhelming the rest of the world is that they are fed upon by other insects.

Insects that eat other insects are considered in two groups known as predators and parasites. Predators are insects (or other animals) that catch and devour other creatures (called the prey), usually killing and consuming them in a single meal. The prey generally is smaller and weaker than the predator. Parasites are forms of living organisms that live on or in the bodies of living organisms (called the hosts) from which they get their food, during at least one stage of their existence. The hosts usually are larger and stronger than the parasites and are not killed promptly but continue to live during a period of close association with the parasite. Predators are typically very active and have long life cycles; parasites are typically sluggish and tend to have very short life cycles.

Insect Control

Insect control is also important to keep the pests from spreading to other crops, and it may help reduce the incidence of disease by killing insect vectors. Insects attacking vegetables can be divided into three categories.

- Soil Insects,
- Chewing Insects,
- Sucking Insects.

Soil Insects

Soil insects include wireworms, white grubs, fire ants, cutworms, seed maggots and the sweet potato weevil. These insects can be damaging because they feed on the roots, stems and tubers of plants. Often soil insects, especially cutworms, are common in uncultivated soil sites that have had grass and weeds growing the previous season. These undisturbed areas often harbor high populations of soil insects. Once seeds or transplants are planted, soil insects are difficult to control and may begin feeding immediately on the crop. There is a real need for producers to inspect fields for soil insects prior to planting. One or two soil insects per square foot of soil can cause serious damage.

Oftentimes soil insects are clumped in a field, that is, they may be in one area and not in another. Low areas or those areas with the most vegetation often hold the most insects. Controlling soil insects is much easier if done prior to planting. Most insecticides for the control of soil insects should be applied 6 weeks before planting and incorporated into the top 6 inches of the soil. Liquid or granular materials may be used. These can either be broadcast or banded in the row. Sometimes producers will apply insecticides at planting. Make sure to read the label for proper rates and application techniques. Some insecticides may interfere with seed germination and should not be placed in the furrow in contact with the seed.

Chewing Insects

Many chewing insects have a complete life cycle. Therefore, depending on species, there may be one or two damaging stages. Grasshoppers have a chewing-type mouthpart but have an incomplete life cycle. Chewing insects include all species of beetles, grasshoppers and moths and butterfly larvae (most often called worms).

Chewing insects damage foliage, stems and fruit. They may become as numerous as to completely defoliate plants. Eggs of most insects are laid on the plant, and the larvae upon hatching begin to feed. Others may invade the crop by "marching in" or by flying into the field.

Control of chewing insects is basically twofold. One, the grower must watch for eggs and small larvae that begin to feed; two, he must watch for the adults and control them when necessary. Control of these insects is important in the early infestation of the plant. Often, the insect after hatching may bore into the fruit or stem and be hidden from pesticide applications.

These insects often become numerous because producers do not begin treatment early enough. It is vital that fields be watched and these insects controlled at the earliest possible moment.

Sometimes a single application timed properly will control a generation. However, repeated applications are needed to control others like the corn earworm in sweet corn.

Sucking Insects

Sucking insects include aphids ("plant lice"), stink bugs, squash bugs, leafhoppers and spider mites. Spider mites are not insects but are just as damaging and numerous as are some insects. Sucking insects have an incomplete life cycle. After hatching from the egg, they may begin to feed and move about on the plant.

They are usually attracted to the most succulent part of the plant. Aphids usually are found in the terminal or on flowers. Stink bugs and squash bugs readily feed on the tender fruit. These insects damage the plant by reducing the vigor or by injecting a toxin or disease-causing organisms into the plant. Heavy feeding may cause flowers to abort or the leaves to turn yellow and fall off. Feeding on the fruit may cause catfacing injury, hard spots or twisted and misshapen fruit.

Control is easiest to obtain soon after the insects hatch from eggs. This is when the insects are the smallest and most vulnerable to the pesticide. Look for egg clusters, so that timing of the insecticide can be more accurate. Most true bugs have large eggs that can be seen without the aid of a magnifying glass. They are often on the undersurface of leaves and laid in tight groups and glued together, or in the case of squash bugs, they may be laid singly but in a loose fitting group and not glued together.

Pest Control with a Minimum of Chemicals

The "if a little is good, more will be better" attitude leads to a serious misuse of pesticides.

Overuse of pesticides has a number of adverse effects:

- Food products may contain unsafe pesticide residues if improperly treated with pesticide.
- Beneficial insects, earthworms and birds may be harmed or killed along with harmful insects if pesticides are carelessly used.
- Each time producers spray they expose themselves to the possibility of inhalation or absorption of the toxin.
- Careless use of pesticides near water may contaminate water supplies.
- Misuse of pesticides can lead to the development of chemical resistance in the target pest.
- The use of pesticides can lead to outbreaks of secondary pest species.

A growing public concern over the use and misuse of pesticides has led increasing numbers of vegetable producers to seek means of natural pest control. Although some people do not have the time or knowledge to practice all available alternative methods for controlling pests, there are many cultural practices which will help reduce losses. Proper soil preparation, careful plant selection and good cultural practices can be combined with biological and mechanical controls to reduce the need for chemical pesticides.

Temperature, humidity, precipitation and natural enemies all influence insect populations. In some years, troublesome insects may not be numerous enough to significantly damage plants. In other years, large insect populations may cause serious damage or completely destroy host plants.

Effective control of specific insects must be preceded by proper identification of these insects. Once an insect's identity is known, you can learn about its life cycle, seasonal cycle, habits and host plants, and thus exercise more effective control measures.

Several control methods are often combined in order to minimize damage by insect pests. Since insect control methods vary in their effectiveness, you may wish to select alternative methods to correspond with differences in plant growth and productivity, insect damage, weather conditions and cultural practices. Various control methods will now be considered.

Resistant Plant Varieties

Use available plant species or varieties which are resistant to, or at least tolerant of, insect activity. Insect resistance in plants frequently is interpreted as meaning "immune to insect damage." Actually, it is a term for distinguishing plant varieties which exhibit less insect damage when compared to other varieties under similar growing and pest population conditions. Some varieties may be less "tasty" to insect pests, or may possess certain physical or chemical properties which discourage insect feeding or egg-laying, or may be able to support large insect populations without suffering appreciable damage.

Before buying seeds or plants, check with your local county Extension agent for information on resistant varieties which will grow well in your area. Examples of vegetable varieties that have shown resistance to specific insect pests are listed in table below. Some varieties may be resistant to insect attack, but may be subject to certain restrictions such as soil pH, drainage or temperature.

Vegetable varieties that have shown some resistance to specific insect pests:

Vegetable	Variety	Insect resistance
Bean (snap)	Wade	Striped Flea Beetle
Broccoli	De Cicco	Striped Flea Beetle
Cabbage	Early Globe Red Acre Round Dutch	Cabbage Looper, Imported Cabbageworm Cabbage Looper, Imported Cabbageworm Cabbage Looper, Imported Cabbageworm
Cabbage (Chinese)	Michihili	Diamondback Moth
Collard	Georgia	Striped Flea Beetle, Harlequin Bug
Corn (sweet)	Golden Security	Corn Earworm
Cucumber	Ashley Piccadilly Poinsett	Pickleworm, Spotted Cucumber Beetle Pickleworm Spotted Cucumber Beetle
Kale	Vates	Diamondback Moth
Mustard	Florida Broadleaf	Diamondback Moth, Striped Flea Beetle
Radish	Cherry Belle White Icicle	Diamondback Moth, Harlequin Bug Harlequin Bug
Squash	Early Prolific Straightneck White Bush Scallop Zucchini	Pickleworm, Striped Cucumber Beetle Pickleworm, Striped Cucumber Beetle Striped Cucumber Beetle

Sweetpotato	Centennial Jewel	Sweetpotato Flea Beetle, Potato Wireworm Sweetpotato Flea Beetle, Potato Wireworm
Turnip	Seven Top	Diamondback Moth, Striped Flea Beetle
Rutabaga	American Purple Top	Diamondback Moth, Striped Flea Beetle

Cultural Controls

Many cultural practices can be used to reduce the potential for, or actual damage of plants caused by insects:

- Plowing and cultivating exposes soil insects to adverse weather conditions, birds and other predators. In addition, deep plowing will bury some insects and prevent their emergence.
- Crop rotation can be effective against insects that develop on a narrow range of food plants and also against insects with short migration ranges. Movement of crops to different sites will isolate such pests from their food source. If an alternate site is not available, then change the sequence of plants grown in the field. Do not plant members of the same plant family in the same location in consecutive seasons. For example, do not follow melons with cucumbers or squash.
- Proper use of fertilizers and water will induce healthy plant growth and increase the capability of plants to tolerate insect damage. However, excessive amounts of organic matter or manure can encourage millipedes, pillbugs, white grubs and certain other pests.
- Changes in planting or harvesting time often will reduce plant damage or keep insect pests separated for susceptible stages of the host plant. Delayed planting, until the soil is warm enough for corn and bean seeds to germinate quickly, reduces seed maggot damage. Hot caps or row covers placed over plants used during the early season not only will preserve heat, but also will protect plants from damaging wind, hail and insects. In some situations, a healthy transplant will overcome insect damage more easily than a small plant developing from seed in the field.
- Removing soil residues and disposing of weeds and other volunteer plants eliminates food and shelter for many insect pests such as cutworms, webworms, aphids, white grubs, millipedes and spider mites. When plants stop producing, till them into the soil or take them to the compost pile.
- Companion planting (an orderly mixing of crop plants) is a cultural practice aimed at diversifying insect populations. Numerous claims have been made about the ability of certain plants to protect certain other plants from insect damage. However, no data from scientific studies are available to prove the value of companion plantings.

Mechanical Control Methods

Preventive devices often are easy to use, although their effectiveness varies. Here are examples of such devices:

- Cheesecloth or spun bound polyester row covers for plant beds, hot beds and cold frames to prevent insect egg laying.

- Mesh covers for tomatoes and other plants to keep out large insects and birds.
- Aluminum foil mulch to repel aphids.

Black light traps are effective tools for monitoring insect species in a given area, but usually provide little protection for the crop. Light traps attract both harmful and beneficial insects that ordinarily would not be found in the area. Attracted insects may not be caught in the traps, but may remain in the area, and the harmful ones may cause damage later. Also, some species such as wingless insects and those insects only active in the daytime are not caught in the traps. Consequently, the value of black light traps is questionable. Where black lights are used, it is recommended that they be placed 50 to 75 feet away from the area which is to be protected.

Biological Control Methods

Generally, biological control can be defined as the direct or indirect use of parasites, predators or pathogens (bacteria, viruses, fungi protozoans) to hold pest insect populations at low levels to avoid economic losses. Biological control methods fall into three categories:

- Introduction of natural enemies which are not native to the area (these enemies must then establish and perpetuate themselves).
- Enlarging existing populations of natural enemies by collecting, rearing and then releasing them back into the environment.
- Conservation of beneficial organisms by such means as the judicious use of pesticides and the maintenance of alternate host insects, so parasites and predators can continue to develop.

Many beneficial organisms occur naturally around crops, but often they are not numerous enough to control a pest before it inflicts severe damage. In fact, parasites and predators appear to be most effective when a pest population has stabilized or is relatively low. Their influence on an increasing pest population usually is minimal since any increase in parasite and predator numbers depends on an even greater increase in pest numbers. Pathogens, however, seem to be most effective when pest populations are large. Consequently, the nature of the host insect-natural enemy relationship makes it impossible to have an insect-free environment and at the same time maintain sizable populations of beneficial insects.

The following is a list of some of the more popular biocontrol agents:

- Bacillus thuringiensis: (Dipel, Thuricide, Biological Worm Killer). This bacterial insecticide provides effective control of the larvae of several moths and butterflies. The bacterial spores are harmless to warm blooded animals and beneficial insects.
- Bacillus popilliae: (Milky Spore, Doom, Japedimic). This bacterial insecticide controls grubs of Japanese beetles in the eastern U.S., and some testing has been done for control of white grubs (Phyllophaga spp. and Cotinis spp.) in Texas. It has not been effective against the principal white grub species in Texas.
- Nosema locustae: A spore (Protozoan) used to control grasshoppers. The material is sprayed on the plants which grasshoppers ingest. The spores germinate inside the grasshopper,

causing death. Control is extremely slow and growers may not be satisfied with results. Baits have proven more effective.

- Trichogramma wasp: Adult wasps are available from several sources. The tiny wasps attack the eggs of more than 200 pest species, including cutworms, armyworms, fruit worms and many moth and butterfly eggs deposited in orchards and field crops. Wasps should be released when the moths are first seen, but a sequence of releases throughout the season is preferable to a single, large release. Results will depend on the timing of the releases, selection of Trichogramma species and placement of wasps near host egg masses.
- Green lacewings (Chrysopa): The larvae, known as aphid lions, prey on many vegetable pests including aphids, spider mites, leafhoppers, thrips, moth eggs and small larvae. Adult lacewings feed on honeydew, nectar and pollen. Introduced lacewings must have a readily available supply of food, or they will leave. Eggs are sometimes available.
- Praying Mantis: Egg cases, containing about 200 individual eggs, are available from a number of sources. The mantis is a voracious predator. In addition, it is cannibalistic immediately after hatching, so few nymphs survive the first week of life. However, the mantis is a poor searcher for food and usually waits for prey to come to it. This greatly influences the kinds of insects it captures and kills. Food preferences include grasshoppers, crickets, bees, wasps and flies.
- Lady beetles: Adult beetles are available from several sources. Aphids are the preferred hosts, but lady beetles will eat mealy bugs, spider mites and certain other soft bodied pests and eggs. They do not, however, kill grubs, caterpillars and other beetles. Unless an ample supply of live aphids or other hosts are available at the release point, lady beetles will disperse and leave the area. In some cases, most of the beetles will leave the area regardless of the availability of food. Lady beetles can be encouraged to remain on a plant by using small meshed screen cages (remove cages before they devour all of their food supply). In hot dry weather these beetles enter a nonactive (aestivation) state in which feeding and reproduction cease. Little control from lady beetles can be expected during this time.

Natural Insecticides

Despite all efforts, noninsecticidal methods at times will fail to prevent excessive insect damage. At such times, the use of insecticides may be the only alternative left. Insecticides chosen should have only low toxicity for humans and other warm blooded animals. They should be used only when needed and according to label directions. A better understanding of insecticides will enable you to use these materials more effectively and to realize that they can be an aid without harming you or the environment.

- Pyrethrum: Botanical Insecticide. This slightly toxic insecticide is derived from the flowers of a species of Chrysanthemum imported mainly from Kenya and Ecuador. The material causes rapid paralysis of most insects, but the insects usually recover unless the pyrethrum is used in combination with a synergist or other poison. Pyrethrum, mixed with synergists such as piperonyl butoxide or piperonyl cyclonene to increase toxicity and produce longer residual action, is used extensively in crop sprays and dusts. This chemical is registered for use on most vegetables at any time during the growing season.

- Nicotine: Botanical Insecticide. Pure nicotine is a tobacco extract highly toxic to warm blooded animals. The insecticide usually is marketed as a 40% liquid concentrate of nicotine sulfate, which is diluted in water and applied as a spray. Dusts can irritate the skin. Nicotine is used primarily for piercing-sucking insects such as aphids, whiteflies, leafhoppers and thrips. Nicotine is more effective when applied during warm weather. It degrades quickly, so it can be used on many food plants nearing harvest. It is registered for use on a wide range of vegetable and fruit crops.
- Sabadilla Botanical Insecticide: Sabadilla is obtained from the seeds of a lily-like plant and acts as both a contact and stomach poison for insects. It is not particularly toxic to mammals, but does cause irritation of the eyes and respiratory tract. A mask should be worn when working with this insecticide. This material deteriorates rapidly upon exposure to light and can be used safely on food crops shortly before harvest. Sabadilla generally is used as a 5 to 20% dust or as a spray.
- Rotenone: Botanical Insecticide. Rotenone is extracted from the roots of Derris plants in Asia and cube plants in South America. This general insecticide is harmless to plants, highly toxic to fish and many insects, moderately toxic to mammals, and it leaves no harmful residues on vegetable crops. It acts as both a contact and stomach poison to insects. It is slow acting, and in the presence of sun and air, its effectiveness is lost within a week after application. Wear a mask during application because rotenone can irritate the respiratory tract. Rotenone dusts and sprays have been used for years to control aphids, certain beetles and caterpillars on plants.

Organic Controls

Growers have been using soap to control insects since the early 1800's. Researchers have not yet determined exactly how soaps work. Some soap simply wash off the outer waxy coating of the insect cuticle, destroying its watertight nature and causing the insect to dry up and die. Other soaps have additional insecticidal properties which may affect the nervous system. These soaps appear to have toxic activity only against plant eating insects, and thus may spare beneficial insects such as lady beetles, honeybees, lacewings and predatory mites. Although a number of soaps tested have insecticidal properties, only Safer's Insecticidal Soap is currently registered for use on edible crops. It controls such pests as spider mites, aphids, mealybugs, whiteflies, harlequin bugs, stink bugs and thrips.

Organic growers have been using a spray mixture containing onions, garlic and pepper mixed together to control insects for many years. Research indicates that of combination of these materials have been erratic and in many cases ineffective for insect control. Sprays of food-derived substances do not appear to be good choices as a pesticide. Some success may be achieved with them, but it is likely to be sporadic. Spraying several times a week might help to bring infestations under control. Control with one application should not be expected.

Controlling Insects with Pesticides

Insects often pass the winter in a life stage that can damage crops early the next season. They may pass the winter in the field in grass, trash or in weedy fence rows. Squash bugs spend the winter in the adult stage and often invade fields soon after planting and begin feeding. Good squash bug control can be achieved if producers realize this and treat them before the plants actually begin to bloom. Deep plowing of fields in the winter can eliminate some overwintering stages.

Fall grown vegetables are usually heavily attacked by insects. More insecticides may have to be used on fall vegetables.

Some cultural control measures can be used to suppress insect pressure. Examples of these measures are deep plowing, controlling weeds and grass around the field and destroying crop residue soon after harvest.

Carbaryl (Sevin) insecticide is an excellent material, and vegetables sprayed with Carbaryl can be harvested soon after spraying. However, repeated use of Carbaryl often results in a spider mite outbreak.

Safe use of Pesticides

When it is necessary to use insecticides, use them wisely and safely. The following tips will help you make better use of insecticides:

- Inspect plants and monitor insect numbers and activity on a regular basis. Pay particular attention to underside of leaves where insects and their eggs frequently occur. If treatments are applied when an infestation first starts, insect numbers can be maintained at lower levels much more easily and with smaller amounts of chemicals.
- When applying insecticides to plants, treat all plant surfaces unless otherwise stated on the product's label. This ensures that an insect anywhere on the plant will be exposed to a lethal amount of the chemical. Do not apply insecticides to wilted plants or during the hottest part of the day. Apply dusts only when the wind is calm and plants are dry. Sprays should be applied when the wind is no more than 5 to 10 mph. Retreatment may be necessary after rainfall.

Common Insects Attacking Vegetable Crops

Soil Insects

- Wireworm

  Plants Attacked: Corn, small grains, grasses, potatoes and other root crops such as sweet potatoes.

  Description: Adults are usually hard shelled, brownish gray or nearly black, somewhat elongated with the body tapering toward each end. The larvae usually are hard, dark brown, smooth, wire like worms from ½ to 1½ inches long when grown.

  Life History: Young adults remain in the soil until spring. The subsequent egg stage requires a few days to a few weeks to hatch. Larvae spend from two to six years in the soil feeding on roots of grasses and other plants. Pupation usually is completed in a few weeks.

Damage: Crops may fail to emerge, or stay thin and patchy. Wireworms feed on seed and underground plant parts. On potatoes they cause small holes or trail-like appearance in their feeding habits.

- White Grubs

Plants Attacked: Corn, beans and potato tubers and other root vegetable crops.

Description: Adult beetles are from ½ to 1 inch long, vary from light to dark brown and are robust in form. The larvae are white, curved bodied grubs with brown heads and three pairs of legs. The hind part of the abdomen appears darker because soil particles inside show through the body wall. There are probably 100 species, and many cause damage.

Life History: The life cycle of the more abundant species extends over three years. In late spring pearly white eggs are deposited from 1 to 8 inches deep in the soil. Approximately 3 weeks later the eggs hatch, and the larvae feed on roots and decaying matter. In autumn when cooler temperatures prevail, the larvae emigrate downward and remain inactive until the following spring when they return to feed on plant roots near the soil surface. Greatest damage occurs at this time. At the next autumn, they begin to go deep into the soil, returning to the surface in the spring of the third year; they feed until June. Then oval, earthen cells are made and pupation follows. Adult beetles form in the pupae in a few weeks; they remain in the cells throughout the winter and emerge from the soil the following year to begin feeding, mating and egglaying. In Texas the period from egg to adult seems to be two years for most species.

Damage: Most severe damage by grubs occurs on crops which follow grass sod the next year. Grubs feed on the roots of crops attacked and destroy the root system.

Chewing Insects

- Fruit Worms

There are many types of worms that feed on plants, the fruit or leaves and buds of the plant. These worms include armyworm, beet armyworm, fall armyworm, melon worms, pickle worms, tomato hornworm, tomato pinworm, yellow striped armyworm and the corn ear worm, or the tomato fruit worm.

Plants Attacked: Sweet corn, beans, peas, beets, peppers, melons, squash, cucumber, cauliflower, broccoli and similar crops. The description of these various insect larvae is completely different. Many have elongate bodies, may be striped with spots, and most of them are green to brown to reddish color. The stripes may be white or yellow. Most all of them have three legs behind the head and five sets of legs along the abdomen.

Life History: Many spend the winter as pupae 2 to 6 inches below the soil surface and emerge as moths during the spring and early summer to begin depositing eggs on their favorite crops or host plant. Fresh laid eggs usually are waxy white, but soon turn yellow and darker as the insect matures inside. They are about half the size of a pinhead and variously shaped. A female can deposit from 500 to 3,000 eggs singly on foliage and fruit of many plants. They prefer the tender growing point of the plant to deposit eggs, but eggs commonly are deposited in the curl of various plants and in the whirl of corn or grassy crops. Later in the season, they may deposit eggs directly on the fruit of the plant. The eggs usually hatch in 2 to 4 days, but it may take up to ten days in cold weather; the worm stage lasts 2 to 4 weeks. Full grown larvae crawl down the host plant or drop to the ground when they burrow forming a wall shell cell and pupate. The adult moths usually emerge in 10 to 25 days following pupation. Time from egg to adult varies from 1 to 2 months depending upon weather conditions. There may be 4 to 7 generations of these various worms throughout the year.

Damage: Newly hatched larvae begin feeding immediately on the part of the plant where the eggs were laid. They may then bore into the fruit such as tomatoes or the ears of corn or feed on pods of beans and peas. The worms cause considerable damage because they gnaw or eat out sections of the fruit, and may cause the fruit to rot.

- Cabbage Looper

Plants attacked: Cabbage, cauliflower, broccoli, Brussels sprouts, lettuce and occasionally beans, tomatoes and other crops.

Description: Light, grayish brown moth with a small lighter colored spot near center of forewings. Moths have a wing spread of about 1½ inch. Larvae are light green caterpillars with a few white or pale yellow stripes. Larvae travel with a characteristic looping motion. They have three legs behind the head and three legs at the tip of the abdomen.

Life History: There are continuous generations in the Lower Rio Grande Valley with reproduction slowing down during cold periods. In colder areas the insects overwinter as pupae in flimsy silken cocoons attached to plant residue. A complete generation occurs in 3 to 6 weeks.

Damage: Cabbage loopers are voracious feeders which can strip foliage from infested plants in a short time. Often, when cabbage looper populations become crowded, a virus disease strikes causing high larval mortality.

- Thrips

Plants Attacked: General feeders on vegetables, flowers and field crops.

Description: These are slender, spindle shaped, active insects varying from pale yellow to yellowish-brown. Adults average about 1/25 inch long. Four slender wings are present on females, fringed with long hairs and black margins. Males are wingless, and the larvae resemble adults but have no wings and are smaller.

Life History: The minute eggs are inserted into leaves or stems. These hatch in 2 to 10 days. The larval stage lasts from 5 to 30 days. Adult females can reproduce regularly without mating with the rarely found males. All stages can be found during warmer months, but during colder months, only adults and larvae can be found. It is probable that 5 to 8 generations occur per year, but more may occur in the warmer parts of the state.

Damage: Thrips puncture plants, rasp the surface and then suck the juice. This causes the formation of whitish blotches that first appear as dashes. Severely attacked plants develop a gray or silver appearance and may become distorted. Damage may be found first in the leaf sheaths and stems or in the undersides of a bent leaf where the insects always are most abundant.

Sucking Insects

- Aphids

Aphids are small, sluggish soft-bodied insects often called plant lice. Most species give birth to living young and the young build up very rapidly.

Plants Attacked: All vegetables. Most common vegetables are peas, beans, tomatoes, lettuce, turnips, broccoli and corn.

Description: The most common aphids are the melon or cotton aphid, green peach aphid, cabbage aphid and the pea aphid. They are usually 1/16 inch long, soft-bodied and pear-shaped. They may be black, gray, green, red or yellow, depending on the species.

Life History: Most species give birth to living young; they do lay eggs, and some have wings. There can be 15 to 20 generations per year of certain species.

Damage: Aphids congregate in large numbers and, therefore, may infest vegetables such as mustard greens and leaf lettuce; they also may cause the plant to stunt. Disease transmission is probably more important than actual feeding damage.

- Squash Bugs

Plants Attacked: All cucurbits with preference for squash.

Description: The adults are brownish gray to dark gray bugs about 5/8 inch long. The immature, or nymphs, when first hatched are green with black legs. Later, they become grayish-white with nearly black legs and antennae.

Life History: Adults overwinter, unmated in any type of shelter. They appear in the spring as plants begin to vine and mate. Yellowish to bronze-brown eggs are laid in clusters on the underside of leaves usually in vine angles. Eggs hatch in 1to 2 weeks, and nymphs feed in groups on the stems of the plant for 6 to 8 weeks before transforming to adults.

Damage: Leaves attacked by the squash bug will rapidly become black, crisp and dead. Attacked plant stems often are enlarged but later wither and die.

- Stink Bugs

Plants Attacked: Seed beets, okra, squash, beans, peas, corn, cowpeas, and tomatoes.

Description: The adults are approximately ½ inch long, and each has a triangular shaped shield on the back that extends just back of the shoulders narrowing posterior to a point. Front wings are thickened and stiff about the base, but the distal half is much thinner and membranous. Crushed bugs often have an odor fitting their name. The nymphs are without wing covers and smaller but otherwise similar to adults.

Life History: Life history and habits of each of the stink bugs are similar. Generally barrel shaped eggs are deposited in clusters usually on the underside of foliage. Eggs often are beautifully colored and ornamental. Development from egg to adult occurs in 4 to 6 weeks. From 1 to 3 or perhaps four generations may occur annually. They overwinter as adults in places affording protection from cold weather.

Damage: Damage is caused by nymphs and adults sucking sap primarily from pods, buds, blossoms and seeds. In removing the liquid from contents of developing seeds causes them to become flattened and shriveled. If the pods are attacked at an early stage of development, catfacing or pitted holes will occur on bean pods and squash fruit.

- Whiteflies

  Plants Attacked: Potato, tomato, eggplant, pepper and sweet potato.

  Description: Adults are 1/16 inch in length. They have four wings which along with the dorsal part of the body are covered with white, waxy powder. The nymphs are light green, oval, flattened and about the size of a pinhead. They are attached to the leaf surface until mature, with the last instar more elevated and slightly segmented. The bodies are covered with radiating long filamentous threads resembling young, soft scale insects.

  Life History: Overlapping generations occur in the Lower Rio Grande Valley during spring, summer and fall. Adults emerge, mate and begin depositing elongated yellow eggs and attaching them to the host plant by short stalk. Before hatching, the eggs darken; nymphal period is one month.

  Damage: Both nymphs and adults feed by sucking plant juices. Heavy feeding gives plants a mottled appearance, or it causes them to turn yellow and die. The sticky honeydew excreted by the insect often glazes the lower leaves and permits development of black sooty mold on plants, thus detracting from the plant's beauty and cutting down on photosynthesis.

- Spider Mites

  Plants Attacked: Various plants are attacked including tomatoes, eggplants, beans, corn, peas and various cucurbits.

  Description: The two-spotted spider mite has two forms, a green form with a dark spot on each side and the more common which is a reddish form. Some species of mites may be yellow. All mites are very tiny and almost microscopic. They are about 1/60 inch long. Life History: Adult mites lay eggs on leaf undersides and spin webs beneath which eggs hatch and mites feed. Spider mites reproduce rapidly during hot, dry weather.

  Damage: Mites pierce leaf tissue and suck sap in the larval, nymphal and adult stages. Plants attacked begin to lose color, fading from green to yellow and eventually turn red. Heavy infestation may kill some plants, and heavy webbing may appear on certain plants.

Plant Disease Management

Plant diseases have caused severe losses to humans in several ways. Starvation and uprooting of families resulted from the Irish famine caused by potato late blight (caused by Phytophthora infestans). A valued resource was lost with the virtual elimination of the American chestnut by chestnut blight (caused by Cryphonectria parasitica). And direct economic loss such as the estimated one billion dollars lost in one year to American corn growers from southern corn leaf blight (caused by Cochliobolus maydis, anamorph Bipolaris maydis). Many plant diseases cause less dramatic losses annually throughout the world but collectively constitute sizable losses to farmers and can reduce the aesthetic values of landscape plants and home gardens.

The goal of plant disease management is to reduce the economic and aesthetic damage caused by plant diseases. Traditionally, this has been called plant disease control, but current social and environmental values deem "control" as being absolute and the term too rigid. More multifaceted approaches to disease management, and integrated disease management, have resulted from this shift in attitude, however. Single, often severe, measures, such as pesticide applications, soil fumigation or burning are no longer in common use. Further, disease management procedures are frequently determined by disease forecasting or disease modeling rather than on either a calendar or prescription basis. Disease management might be viewed as proactive whereas disease control is reactive, although it is often difficult to distinquish between the two concepts, especially in the application of specific measures.

Specific management programs for specific diseases are not intended since these will often vary depending on circumstances of the crop, its location, disease severity, regulations and other factors. Most states have some agency such as the Agricultural Extension Service or State Department of Agriculture that formulates and promulgates disease management recommendations for that state. Involvement of these agencies is especially important where the practices include some regulated component such as pesticides or quarantines.

Plant disease management practices rely on anticipating occurrence of disease and attacking vulnerable points in the disease cycle (i.e., weak links in the infection chain). Therefore, correct diagnosis of a disease is necessary to identify the pathogen, which is the real target of any disease management program. A thorough understanding of the disease cycle, including climatic and other environmental factors that influence the cycle, and cultural requirements of the host plant, are essential to effective management of any disease.

The many strategies, tactics and techniques used in disease management can be grouped under one or more very broad principles of action. Differences between these principles often are not clear. The simplest system consists of two principles, prevention (prophylaxis in some early writings) and therapy (treatment or cure).

The first principle (prevention) includes disease management tactics applied before infection (i.e., the plant is protected from disease), the second principle (therapy or curative action) functions with any measure applied after the plant is infected (i.e., the plant is treated for the disease). An example of the first principle is enforcement of quarantines to prevent introduction of a disease agent (pathogen) into a region where it does not occur.

The second principle is illustrated by heat or chemical treatment of vegetative material such as bulbs, corms, and woody cuttings to eliminate fungi, bacteria, nematodes or viruses that are established within the plant material. Chemotherapy is the application of chemicals to an infected or diseased plant that stops (i.e., eradicates) the infection. Although many attempts have been made to utilize chemotherapy, few have been successful. In a few diseases of ornamental or other high value trees, chemotherapy has served as a holding action that must be repeated at intervals of one to several years. For example, antibiotics have been infused into plants to reduce severity of phytoplasma diseases of palms (lethal yellowing) and pears (pear decline) and fungicides have been injected into elms to reduce severity of Dutch elm disease (caused by Ophiostoma ulmi) but in all cases the chemotherapeutant must be reapplied periodically. There also are some "systemic" fungicides such as the sterol biosynthesis inhibiting (SBI) and demethylation inhibiting (DMI) fungicides that diffuse into the plant tissues to some extent and eliminate recently established infections.

Exclusion

This principle is defined as any measure that prevents the introduction of a disease-causing agent (pathogen) into a region, farm, or planting. The basic strategy assumes that most pathogens can travel only short distances without the aid of some other agent such as humans or other vector, and that natural barriers like oceans, deserts, and mountains create obstacles to their natural spread. In many cases pathogens are moved with their host plants or even on nonhost material such as soil, packing material or shipping containers. Unfortunately, exclusion measures usually only delay the entry of a pathogen, although exclusion may provide time to plan how to manage the pathogen when it ultimately arrives. Karnal bunt (caused by Tilletia indica) of wheat is an example of a pathogen originally from India that was anticipated.

An important and practical strategy for excluding pathogens is to produce pathogen-free seed or planting stock through certification programs for seeds and vegetatively propagated plant materials such as potatoes, grapes, tree fruits, etc. These programs utilize technologies that include isolation of production areas, field inspections, and removal of suspect plants to produce and maintain pathogen-free stocks. Planting stock that is freed of pathogens can be increased by tissue culture and micropropagation techniques as well as be maintained in protective enclosures such as screenhouses to exclude pathogens and their vectors. Exclusion may be accomplished by something as simple as cleaning farming equipment to remove contaminated debris and soil that can harbor pathogens such as Verticillium, nematodes or other soilborne organisms and prevent their introduction into non-infested fields.

Eradication

This principle aims at eliminating a pathogen after it is introduced into an area but before it has become well established or widely spread. It can be applied to individual plants, seed lots, fields or regions but generally is not effective over large geographic areas.

Eradication of the golden nematode involved removing infested soil, fumigating soil in infested fields and eventually abandoning infested potato fields for housing developments and other uses. Citrus canker eradication involved widespread removal and burning of diseased trees and, in some cases, destruction of entire citrus groves and nurseries. The disease appeared to be contained and the pathogen eradicated, but the disease has reappeared and new attempts at eradication are ongoing.

Eradication can also be on a more modest scale such as the removal of apple or pear branches infected by the fire blight bacterium (Erwinia amylovora) or pruning to remove blister rust cankers (caused by Cronartium ribicola) on white pine branches. Or, it can be the sorting and removal of diseased flower bulbs, corms or rhizomes. Hot water seed-treatment of cereal seeds to kill smut mycelium in the seed and heat treatment to eliminate viruses from fruit tree budwood for grafting are other examples of pathogen eradication.

Two programs that are actually forms of protection and not pathogen eradication are barberry eradication for reducing stem rust (caused by Puccinia graminis) of wheat and Ribes eradication for preventing white pine blister rust. The strategy is that removing these alternate hosts breaks the disease cycles and prevents infection of the economically more valuable host. These two examples are mentioned here because they are frequently cited as eradication measures. However, stem rust can readily spread from wheat to wheat in many regions by the uredinial stage although elimination of the aecial host, barberry, may deter or diminish the development of pathogenic races of the rust. The white pine blister rust fungus is perennial in the pine host and eradication of the alternate host only protects noninfected trees but does not necessarily eliminate the pathogen from the area.

Eradication may also be accomplished by destroying weeds that are reservoirs of various pathogens or their insect vectors. Elimination of potato cull piles is an effective method of eradicating overwintering inoculum of the late blight pathogen.

Soil fumigation has been a widely used eradication strategy. This technology involves introducing gas-forming chemicals such as carbon disulfide, methyl bromide, or chloropicrin into soil to kill target pathogens. However, undesirable side effects such as killing beneficial organisms, contamination of groundwater, and toxicity of these chemicals have resulted in less reliance on this approach for disease management. Volatile fumigants like methyl bromide are injected into soil and sealed with a plastic film. Some water-soluble fumigants like metam-sodium can be injected into the soil and the soil simply compacted to form a seal.

Crop rotation is a frequently used strategy to reduce the quantity of a pathogen, usually soil-borne organisms, in a cropping area. Take-all of wheat (caused by Gaeumannomyces graminis) and soybean cyst nematode (Heterodera glycines) are two examples of soilborne diseases that are easily managed by short rotations of 1 and 2 years, respectively, out of susceptible crops, which may include susceptible weed hosts such as grasses in the case of take-all.

Burning is an effective means of eradicating pathogens and is often required by law to dispose of diseased elm trees affected by Dutch elm disease (DED), citrus trees infected by citrus canker or of bean fields infected by halo blight bacteria (Pseudomonas syringae pv. phaseolicola). Propane flaming can effectively destroy Verticillium microsclerotia in mint stems, and flaming potato stems prior to harvest may prevent tuber infection by the late blight pathogen. However, burning agricultural fields is controversial because the smoke creates human health and safety and environmental concerns.

Protection

This principle depends on establishing a barrier between the pathogen and the host plant or the susceptible part of the host plant. It is usually thought of as a chemical barrier, e.g., a fungicide, bactericide or nematicide, but it can also be a physical, spatial, or temporal barrier. The specific strategies employed assume that pathogens are present and that infection will occur without the intervention of protective measures. For example, bananas are covered with plastic sleeves as soon as the fruit are set to protect the fruit from various pests including fruit decay fungi.

Protection often involves some cultural practice that modifies the environment, such as tillage, drainage, irrigation, or altering soil pH. It may also involve changing date or depth of seeding, plant spacing, pruning and thinning, or other practices that allow plants to escape infection or reduce severity of disease. Raising planting beds to assure good soil water drainage is an example of cultural management of plant diseases such as root and stem rots.

Fungicides have been used for more than a hundred years and new fungicides continue to be developed. Bordeaux mixture, a basic copper sulfate fungicide, was the first widely used fungicide and is still used today in various forms. The earliest fungicides were simple elements like sulfur or metallic compounds of copper or mercury, and these are generally classed as inorganic fungicides.

One liability of these recent narrow-range fungicides is that they often have single-site modes of action, (that is, their site-specific activity is controlled by one or a few genes), and thus are especially prone to development of fungicide resistance in the pathogen. Several management strategies have been developed to combat fungicide resistance. These include using mixtures of single-site and multi-site fungicides, alternating applications of fungicides with different modes of action, applying fungicides only when needed instead of on either a calendar or prescription basis, and applying the recommended dosage and not attempting to cut costs by reducing the recommended amount of fungicide applied.

Fungicides can be applied by any of several methods: ground sprayers, airplanes or through irrigation systems, but to be effective applications must be done properly. First, the fungicide must be legally registered for use on the plant involved and against the target disease. Several different chemicals may be registered for the same crop or disease. If the different fungicides are similar in effectiveness, cost, ease of application, and safety, then timing of application becomes the most critical factor. If applied too early much of the chemical will be wasted before it can be effective; if applied too late, it will be largely ineffective. The benefits of properly applied fungicides can often be striking. Distribution of the spray droplets is important; the finer the spray the more complete the coverage on the plant surface. However, very small droplets form a mist that is easily displaced by wind.

Many cultural practices can be modified to manage the occurrence, intensity or severity of plant diseases. These include selection of suitable growing sites for the crop, adequate tillage to bury pathogen-infested plant residues, rotation to nonsusceptible crops, selecting pathogen-free planting stocks, orientation of plantings to improve exposure to sun and air currents, pruning and thinning to eliminate sources of infection and improve aeration in and around susceptible plants, water management on both plants and in soil, adequate nutrition, proper cultivation to improve root growth and avoid plant injury, and sanitation procedures to eliminate sources of inoculum.

Biological control involves the use of one living organism to control another, and this management technology has received much attention in recent times. However, the number of biological agents registered for use is relatively small, success has been limited, and application has been largely restricted to intensively managed, high value crops such as greenhouse plants. Two examples of effective biological control are the use of the fungus Peniophora gigantea to inoculate tree stumps to prevent infection of adjacent trees by the wood decay fungus Heterobasidion annosum, and the application of the nonpathogenic (i.e., non-tumor-producing) bacterium Agrobacterium radiobacter to fruit trees before planting to prevent infection by the crown gall bacterium (Agrobacterium tumefaciens).

Resistance

Use of disease-resistant plants is the ideal method to manage plant diseases, if plants of satisfactory quality and adapted to the growing region with adequate levels of durable resistance are available. The use of disease-resistant plants eliminates the need for additional efforts to reduce disease losses unless other diseases are additionally present. Resistant plants are usually derived by standard breeding procedures of selection and/or hybridization. A few disease-resistant lines have been obtained by inducing mutations with x-rays or chemicals. There is also interest in chemicals called "plant activators" that induce plant defense responses called systemic acquired resistance (SAR) and induced resistance. Recently, resistant plants have been developed through the use of genetic engineering (e.g., resistance to the Papaya ringspot virus).

Selection of resistant plants involves subjecting plants to high levels of disease pressure and using the surviving plants as sources of disease resistance. Plants that survive this pressure often have genetic resistance that can be utilized directly by propagation or as sources of resistance to develop resistant plants that also have the requisite qualities for that crop. Hybridization is a tactic where a plant having the desired agronomic or horticultural qualities, but is susceptible to a disease, is crossed with a plant that is resistant but which may or may not have the other desirable characteristics such as size, yield, flavor, aesthetics, etc.

Disease escape occurs when susceptible plants do not become diseased for some reason. This may be due to some anatomical or physical character, such as the occurrence of leaf hairs, thick cuticle, or modified stomata, or they may be environmental, in which conditions are not conducive to disease development. Although disease escape based on some anatomical feature is useful occasionally, escape more often complicates the process of developing disease resistant plants.

Development of disease-resistant plants has been relatively successful with annual and biennial plants, but less so with perennials, primarily because of the longer time required to develop and test the progeny. Woody perennials, such as ornamental, forest, and orchard trees, have been especially difficult for plant breeders to develop useful disease resistance. For example, chestnut blight and Dutch elm disease have devastated two valued native trees. In both cases there have been extensive attempts to develop resistant trees, usually by creating hybrids with exotic chestnut or elm trees, and some resistant selections have resulted. Unfortunately, these generally lack the desirable qualities, such as nut flavor or tree forms characteristic of the native trees. Another introduced disease that has impacted native trees is white pine blister rust.

Development of resistance has been most successful against the more specialized pathogens such as rust fungi, smut fungi, powdery mildew fungi, and viruses, but less so against general pathogens such as many blight, canker, root rot and leaf spotting pathogens.

A major problem with genetically resistant plants is that host-differentiated pathogenic races can be selected, so that many breeding programs become continuous processes to develop disease resistant plant lines. Disease resistance conferred by a single major gene is sometimes called specific or qualitative resistance and is race-specific. This type of resistance is often unstable, and emergence of a pathogenic race that can attack that genotype can completely overcome this type of resistance. Quantitative resistance or general resistance derives from many different genes for resistance with additive effects to provide more stable (or durable) resistance to pathogens.

There are several strategies to minimize this race development and resistance failure. These include methods of gene deployment, where different genetic plant types are interspersed on a regional basis to avoid a genetic monoculture, or planting mixtures of cultivars having different genetic compositions to ensure that some component of the crop will be resistant to the disease.

Integrated Plant Disease Management

In a disease pyramid, three principal components are considered. These are host, pathogen, and the environment. All the three components act together during disease development. Thus, to protect a plant from disease(s) one has to tackle all the three parameters at the same time.

The management programme should cover the integration of all the methods; those favour the host, discourage the pathogen and modify the environment. This type of management programme to control plant diseases is called Integrated Disease Management i.e., IDM or Integrated Pest Management i.e., IPM.

Procedure of Integrated Plant Disease Management

The procedure of management is different in respect of the principal component:

Management of Host

It includes:

1. Procedure to improve plant vigour.
2. Induction of disease resistance.
3. Breeding for disease resistance.

Management of Pathogen

It includes:

1. Eradication or reduction of inoculum.
2. Application of chemicals on plant surface, to discourage the pathogen.
3. Prevention through legislation (i.e., keeping the pathogen away from the host) by quarantine, etc.

Management of Environment

It includes:

1. Crop management.
2. Soil management.
3. Water management.

Programmes of Integrated Plant Disease Management:

The programme of Integrated Plant Disease Control can be taken against a particular common disease, such as late blight disease of potato or against all diseases affecting a particular crop such, as Potato, Apple, Citrus etc.

Integrated Management in an Annual Crop

The following programmes must be undertaken for annual plants like potato against different diseases like late blight (fungus), brown rot and ring rot (bacteria), leaf roll of potato (virus – potato virus-1), etc:

1. Stock tuber should be free from any disease i.e., a healthy tuber is needed.
2. Field should be cleaned from previous years plant debris and tubers, if any, where pathogen may perpetuate for the next season.

3. Crop rotation must be done with legume or any other crop plant, except Solanaceous members.

4. Plant should be protected from disease through foliar spray at regular intervals.

5. Make the soil loose by ploughing and expose it to sunlight for few days in summer months and/or fumigate the soil with chemicals.

Integrated Management in a Perennial Crop

For perennial crop like orchard crop (peach, apple, pear etc.); following programme should be undertaken:

1. Nursery-stock should be fumigated (to eradicate the nematode, if present).

2. The stock should be fumigated (to eradicate the nematode, if present).

3. The growing field should be fumigated before planting to eradicate any pathogen like Armillaria, Phytophthora, if present.

4. Proper drainage in the field should be maintained.

5. Stock should not grow in old field near the old plants that may carry different pathogens like canker, etc.

6. The field should be properly irrigated, and supplemented by fertilisers.

7. Plants should be sprayed regularly to protect them from different diseases and insect pests.

8. Usually the fruits become susceptible to several fruit rotting and fruit spotting fungi that attack at different stages till harvest or even in storage. So the fruits should be sprayed at 10-14 days' interval. Sometimes insects can damage at any stage, thereby the fungi infect the fruit through wound. Thus, the insecticides should be applied at intervals.

So the integrated control programme includes several control methods like legislation (quarantine, inspection etc.); cultural practices (sanitation, crop rotation etc.); chemical controls (spraying of plants, fumigation of soil, surface sterilisation of planting materials etc.). All the required procedures for a particular crop should be followed to overcome the disease(s) and/or insect pests.

Controlling Plant Disease

Cultural Methods

1. Selection of Geographical area which on the basis of the favourable temperature and humidity requirement for a particular crop but unflavored to the fungi and bacteria.

2. Selection of field: Many soil borne diseases are controlled by proper selection of the field. It is quite possible that a particular field soil contains a pathogen species. In that case the

particular crop is not sown in that field for several years. The causal organism of Red rot of sugarcane Colletotrichum falcatum survives in the soil for several months. Water drainage is also taken care of while selecting the field.

3. Choice of the time of sowing: The susceptible stage of plant growth and the favourable environment for pathogen should not match at the same time.
4. Disease escaping varieties: Certain varieties of crop due to their growth characteristics are able to escape from disease. This disease escaping characteristics of the crop is not genetic rather it is due to growth habits and time of maturation. Early maturing variety of pea is capable of escaping powdery mildew and rust.
5. Selection of seed: To avoid seed borne diseases, healthy and disease free seeds are essential.
6. Crop rotation: Crop rotation is essential for controlling soil borne diseases and pathogens.
7. Roguing: Removal and Destruction of Diseased Plant Organs, eradication of alternate and collateral hosts and sanitation of Fields.
8. Modification of cultural Practices: Cultural practices such as – distance between the plants, time and frequency of irrigation, transplantation time and method, mixed cropping, amount and property of fertilizer and compost etc. can be changed to reduce losses caused by the disease.

Eradication of Insect Vectors

Insects serve as vectors for many diseases. Eradication of such insect vectors is essential for the control of pathogens.

Examples of some diseases and their insect vectors are as follows:

Agar: A gelatin-like substance obtained from sea weed (red algae Gracilaria, Gelidium etc.) and used to prepare culture media on which microorganisms are grown for study.

Alternate host: One of the two kinds of plants on which a parasitic fungus (e.g. black rust of wheat caused by Puccinia graminis trifid) must develop to complete life-cycle.

Anthracnose: A leaf spot or fruit spot type of disease caused by fungi that produce their sexual spores in an acervulus.

Antibiosis: The phenomenon in which a substance produced by one microorganism is harmful to another organism.

Antibody: A protein produced by specific stimulation when a foreign antigen enters into the blood of an organism. Antibodies get attached with the antigens and make them ineffective or harmless.

Antigen: A substance (usually a protein, lipid or carbohydrate) which after entering into a body activates the production of antibody.

Bacteriophage: A virus which infects specific bacteria and kills them.

Bacteriophage: A chemical or physical agent that prevents', multiplication of bacteria without killing them.

Blight: A non-restricted tissue disintegrating symptom characterized by general and rapid killing of leaves, flowers & stem.

Blotch: A disease characterized by large and irregular spot or lesions on leaves, shoots and stems.

Canker: A necrotic or sunken lesion Oil a stem, branch or. fruit of a plant (e.g. citrus canker caused by Xanthomonas citri.

Carries: A plant or an organism which carries an infections agent but does not show symptoms of disease produced by the agent.

Chlorosis: Yellowing of green tissue due to chlorophyll destruction.

Damping off: Destruction of seedling near the soil surface, resulting in the falling of seedling on the ground.

Die Back: Progressive death of shoots and roots generally starting at the tip.

Disinfectant: An agent that kills or inactivates pathogens in the environment or on the surface of the plant, prior to infection.

Downey Mildew: A plant disease in which the mycelium & spores of the fungus appear as a Downey growth on the host surface.

Endemic Disease: A disease which regularly occurs on a particular area of earth or country.

Epidemic Disease: A wide spread & severe outbreak of a disease.

Etiolation: Yellowing of the plant due to deficiency of light.

Exclusion: Control of plant disease by excluding the pathogen or infected plant material from disease free areas.

Exudate: Liquid discharge from plant tissue.

Facultative parasite: An organism that is usually saprophyte under certain conditions may become parasite.

Facultative saprophyte: An organism that is usually parasite but may also live as a saprophyte.

Gall: A swelling produced on a plant as a result of infection by certain pathogens.

Gummosis: Production of gum by plant tissue.

Heterotrophy: An organism depending on an outside source for organic nutrients.

Hyperplasia: Excessive development due to increase in the number of cells.

Hypertrophy: Excessive growth due to increase in size of cells.

Immunity: The state of being exempted from infection by a given pathogen.

Infection: Establishment of the pathogen in the host.

Infections disease: A disease caused by a pathogen which can spread from a diseased to a healthy plant.

Latent virus: A virus that does not induce symptoms in its' host.

Leaf mottling: A disease caused due to Zn deficiency (specially in Citrus spp.) in which new leaves develop inter-veinal chlorosis, get reduced in size, plant becomes bushy and the branches show die-back symptoms.

Lesion: A localized area of discoloured, diseased tissue.

Mildew: A plant disease caused by a fungus in which the mycelium and spores are seen as a whitish growth on the host surface.

Mosaic: Symptom of certain viral diseases of plants characterized by intermingled patches of normal and light green or yellowish colour.

Mycorrhiza: Symbiotic relationship between roots of higher plants and fungal mycelia which is essential for the growth of these plants.

Necrosis: The death of cells or of tissues.

Obligate parasite: A parasite that in nature can grow and multiply only on living organisms.

Parthenogenesis: Formation of embryo without fertilization.

Pathogen: An disease causing agent in plant.

Plasmogamy: Fusion of cytoplasm's of two cells.

Polymorphism: Having various forms in a life cycle. The rust fungus is allomorphic as it produces five different types of spores in its life-cycle.

Pustule: Small blister like elevation of epidermis.

Quarantine: Control of export and import of plant to prevent spread of diseases or pests.

Race: A genetically distinct mating group within a species; also a group of pathogens with distinct pathological or physiological characteristics.

Resistance: The ability of an organism to overcome, completely or partially the effect of a pathogen.

Rickettsia like Organisms: RLOs a prokaryotic microorganism having a cell wall and obligate intra-cellular parasite.

Ring spot: A circular chlorotic area with a green centre; symptom of many viral diseases.

Rot: The softening, discolouration and disintegration of a succulent plant tissue as a result of fungal or bacterial infection.

Russetting: Brownish roughened areas on fruit skin produced as a result of excessive cork formation.

Rust: A disease of grasses and other plants giving a rusty appearance to the plant and caused by uredinales (rust fungi).

Saprophyte: A organism which lives on dead and decaying organic matter.

Scab: A rough, crust like diseased area on the surface of a plant organ. A disease in which such areas are formed.

Scorch: Burning of leaf margins as a result of infection or unfavorable environmental conditions.

Smut: Disease caused by Ustilaginaceae, characterized by masses of dark, powdery spores.

Spot: Disease symptom in which certain restricted tissue.

Susceptibility: The inability of a plant to resist the effect of a pathogen.

Susceptible: A plant or species which is incapable of resisting the effect of a pathogen.

Toxin: A compound produced by microorganisms and being toxic to a plant or animal.

Vector: An insect able to transmit a pathogen.

Vein banding: Bands of green tissue along the veins while the tissue between the veins become chlorotic.

Viroid: A naked nucleic acid which resembles virus but is devoid of protein coat.

Wilt: Loss of rigidity and dropping of plant parts wholly or partially.

Yellows: Yellowing and stunting of host plant.

Physical Methods

The hot water treatment method of Jensen was developed in 1887 which was used to control loose smut disease of wheat, barley and Oats. Until the development of systemic fungicide hot water treatment was the only method to control loose smut. Hot water treatment is also effective in the control of nematodes.

Solar energy treatment to control loose smut was first developed by Lutlzra. In this method seeds are first rinsed or soaked in water for 4-5 hrs. before drying them in scorching sun. Hot air treatment for the control of virus in propagating stocks was first developed by Kunkal in Peach yellow.

Chemical Methods

Soil Treating Chemicals

It is used for controlling such soil borne diseases which attack on seeds or seedlings. The examples of such chemicals are – Formaldehyde, Captan, Thiram, Zineb, Organo-mercurials, PCNB, Ethylene dibromide, vapam etc.

- For Externally seed borne diseases, chemicals such as formalin, copper carbonate, captan, organo-mercurials (Agrosan GN and Ceresan) are used for seed treatment.
- For Internally seed borne diseases (i.e. loose smut), hot water treatment and solar treatment are used.
- Systemic Organic Compounds are effective chemicals for controlling both externally and internally seed borne diseases eg. Oxanthin derivatives (Plantvax and Vitavax), Benlate, Bavistin, Demosan.
- For controlling air borne diseases, foliar application of chemicals is more effective.
- The common copper fungicides are: Perenox, Perelan, Blitox, Cuprokyt, Cuprosanand Fytolan. Its use is comparatively better than that of Bordeaux mixture.

Seed dressing with organomercurials and systemic fungicides.

Plant Quarantine

Plant quarantine can be defined as a legal restriction on the movement of agricultural commodities for the purpose of exclusion, prevention or delay m the establishment of plant pests and diseases in areas where they are not known to occur.

References

- Crop-science-innovations-crop-protection: bayer.com, Retrieved, 31 July, 2019
- Crop-protection-chemicals, materials-and-applications: essentialchemicalindustry.org, Retrieved 1 August, 2019
- Weed, plant: britannica.com, Retrieved, 24 March, 2019
- Cultural-control, management, weeds, invasive, biodiversity: environment.gov.au, Retrieved, 17 February, 2019
- Insect-management, texas-vegetable-growers-handbook, guides, vegetable: aggie-horticulture.tamu.edu, Retrieved, 28 Aprıl, 2019
- Plant diseasemanagement, topc, disimpactmngmnt, edcenter: apsnet.org, Retrieved, 21 July, 2019
- Integrated-plant-disease-management, plant-pathology: biologydiscussion.com, Retrieved, 15 January, 2019
- Top-4-methods-of-controlling, plant-diseases-agriculture, plant-diseases, plants: biologydiscussion.com, Retrieved 18 May, 2019

Permissions

All chapters in this book are published with permission under the Creative Commons Attribution Share Alike License or equivalent. Every chapter published in this book has been scrutinized by our experts. Their significance has been extensively debated. The topics covered herein carry significant information for a comprehensive understanding. They may even be implemented as practical applications or may be referred to as a beginning point for further studies.

We would like to thank the editorial team for lending their expertise to make the book truly unique. They have played a crucial role in the development of this book. Without their invaluable contributions this book wouldn't have been possible. They have made vital efforts to compile up to date information on the varied aspects of this subject to make this book a valuable addition to the collection of many professionals and students.

This book was conceptualized with the vision of imparting up-to-date and integrated information in this field. To ensure the same, a matchless editorial board was set up. Every individual on the board went through rigorous rounds of assessment to prove their worth. After which they invested a large part of their time researching and compiling the most relevant data for our readers.

The editorial board has been involved in producing this book since its inception. They have spent rigorous hours researching and exploring the diverse topics which have resulted in the successful publishing of this book. They have passed on their knowledge of decades through this book. To expedite this challenging task, the publisher supported the team at every step. A small team of assistant editors was also appointed to further simplify the editing procedure and attain best results for the readers.

Apart from the editorial board, the designing team has also invested a significant amount of their time in understanding the subject and creating the most relevant covers. They scrutinized every image to scout for the most suitable representation of the subject and create an appropriate cover for the book.

The publishing team has been an ardent support to the editorial, designing and production team. Their endless efforts to recruit the best for this project, has resulted in the accomplishment of this book. They are a veteran in the field of academics and their pool of knowledge is as vast as their experience in printing. Their expertise and guidance has proved useful at every step. Their uncompromising quality standards have made this book an exceptional effort. Their encouragement from time to time has been an inspiration for everyone.

The publisher and the editorial board hope that this book will prove to be a valuable piece of knowledge for students, practitioners and scholars across the globe.

Index

A

Agricultural Law, 5-6
Alfalfa, 9, 59, 61, 112, 138
Amino Acids, 3, 37, 67, 168-169
Ammonium Sulfate, 29, 100, 118
Armyworms, 188, 196

B

Bacillus Thuringiensis, 64, 144-146, 195
Beauveria Bassiana, 144, 146
Biomass, 5, 68, 115, 117-118, 123

C

Cabbageworms, 188
Cactus Moth, 186
Carbaryl, 134, 145, 198
Carpel, 16, 53
Cash Crop Farming, 14-15, 86
Cassava, 50, 82, 85-86, 130
Cation Exchange Capacity, 117, 126
Check Row Planting, 90-91
Colorado Potato Beetle, 140-141, 144-146, 188
Corn Rootworms, 63-64, 130, 134
Crop Cultivation, 1, 5
Crop Rotation, 2, 11, 30, 61, 112, 146, 184, 194, 206, 212-213
Crop Selection, 81, 85-87
Crop Yields, 2, 6, 47, 109, 116, 124

D

Deep Plowing, 194, 198
Dibbling, 90
Dry Matter, 70, 82, 101

E

Ethanol, 8, 51, 68-70
European Corn Borer, 53, 64

F

Fall Armyworm, 64, 199
Fiber Crops, 8-9
Forage Crops, 93, 183
Frost Protection, 20, 34, 94

G

Genetically Modified Organisms, 12, 185
Green Lacewings, 196
Green Revolution, 3, 47-49
Gros Michel Banana, 18

H

High-yielding Varieties, 47
Hill Dropping, 90-91
Horn Fly, 149-151

I

Integrated Pest Management, 133, 149, 163, 210

L

Larkspur, 16
Leaching, 20, 29, 106, 108, 115-116, 121, 123, 126, 174
Leaf Curling, 188
Locust Swarms, 131-133, 161

M

Maize Mutant, 53, 55
Manuring, 88-89
Megasporophyll, 16
Molybdenum, 20, 101, 103, 109, 116, 120
Monocropping, 15
Mulching, 146, 182, 184

N

Nitrogen Fertilizers, 100, 103, 106-107, 122
No-till Farming, 61

O

Oil Crops, 8, 10
Ornamental Crops, 8, 10
Oryza Sativa, 40, 82-83

P

Parthenocarpy, 15-16
Parviglumis, 57-58
Perennial Plant, 17
Pillbugs, 194
Ploughing, 82, 88, 212
Poaceae, 36, 72